Collaborative Learning Manual

Collaborative Learning Manual

Vickie J. Aldrich

Doña Ana Branch Community College

Elaine W. Cohen

New Mexico State University and Visioneering Research Laboratory

Lynne M. Hartsell

Doña Ana Branch Community College

ADDISON-WESLEY PUBLISHING COMPANY
Reading, Massachusetts • Menlo Park, California • New York
Don Mills, Ontario • Wokingham, England • Amsterdam • Bonn
Sydney • Singapore • Tokyo • Madrid • San Juan • Milan • Paris

Copyright © 1995 Addison-Wesley Publishing Company, Inc.

ISBN 0-201-59930-9

1 2 3 4 5 6 7 8 9 10 VG 97969594

Preface

Our Goal

Our goal in this book is to provide materials and information for instructors of developmental mathematics courses, to enable them to use collaborative and hands-on learning experiences in their classes. It is intended to be used in conjunction with a standard textbook or other instructional materials, and not as a single text, for the courses presented here. We have included the following aids to help achieve our goal:

- An introductory chapter of suggestions for creating groups and using collaborative learning. Included in this chapter are materials to use throughout the semester to help the students understand the value of group work. We would strongly recommend that you look at this material, especially if you have never used group work in your classes before.

- Notes to the Instructor for each activity that give tips on how to use the activity; tips for grading the activity; and comments on our experiences, or the experiences of others, as we have used the activity. It is important that you look at the notes before using the activities. There is valuable information in the notes that will help you use the activity more effectively.

- Activities for three levels: Basic Mathematics, Introductory Algebra, and Intermediate Algebra. A chapter of activities is included for each level, but some activities for one level may be appropriate in one or both of the other levels. The level of mathematical sophistication required for the student to complete each activity increases as you progress from Basic Mathematics to Intermediate Algebra.

- A range of types of activities, some requiring in-class time only and others requiring outside group work, either for as long as a few days or as much as two weeks. Some activities are presented in detailed scenarios, some are presented as games, and others are a collection of thought-provoking questions. Each will appeal to the adult student in some way, though not all will appeal to every student. Also included are open-ended problems to help foster critical thinking.

Why Use Groups?

The *Curriculum and Evaluation Standards for School Mathematics*, published in 1989 by the National Council of Teachers of Mathematics (NCTM), lists a summary of "mathematical expectations for new employees in industry" for the year 2000 and beyond, as formulated by Henry Pollack, a noted industrial mathematician. The list includes:

- The ability to work with others on problems.

- The ability to see the applicability of mathematical ideas to common and complex problems.

- Preparation for open problem situations, since most real problems are not well formulated.

NCTM then makes the following recommendation:

> To develop such abilities, students need to work on problems that may take hours, days, and even weeks to solve. Although some may be relatively simple exercises to be accomplished independently, others should involve small groups or an entire class working cooperatively. Some problems also should be open-ended with no right answer, and others need to be formulated.

The United States Department of Labor has also put out its recommendations of what is needed for "solid job performance." One of the five competencies listed is:

> Interpersonal Skills—working on teams, teaching others, serving customers, leading, negotiating, and working well with people from culturally diverse backgrounds.

Collaborative learning helps achieve this goal by allowing students opportunities to

- work with their fellow classmates on teams

- participate in peer teaching

- practice leading, negotiating, and working well with people who may come from different backgrounds.

These skills can then later be transferred to the workplace in an effective way.

The Activities

In response to the preceding recommendations, and our experiences as teachers with students who need guidance in making mathematics a part of their everyday lives, we have written this manual. The activities in this *Collaborative Learning Manual* are designed to help students:

- Work together in groups to solve problems that stimulate group participation and interaction. By working in groups, they will be better prepared for employment and will have the benefits of peer teaching.

- Get a feel for how mathematics can be useful in their everyday lives by working on "real-life" problems. For students, many problems in textbooks appear fabricated to work out perfectly and have nothing to do with what they will encounter in their lives. Presenting problems in a detailed scenario or that are open-ended gives students the feel of being from the world outside academia.

- Think about mathematics more clearly by having to write out completed assignments in English, including their thought processes. Students will consolidate and internalize the process used to solve problems by having to write their thoughts down clearly and concisely.

- See how concrete objects are related to the abstract concepts taught in mathematics courses by using manipulatives in the learning process. The manipulatives need not be specially designed objects, but can be everyday items.

- Find that mathematics can be fun and interesting rather than painful and frightening. Students may then be able to build confidence and find that mathematics really does help them in their lives.

Acknowledgments

Many people have helped in the development and editing of this book. We would first like to thank our families, Bob Hartsell, Marcus Cohen, Jenny Stuart, and Mikaela Cohen, for their patience and support. We would also like to thank the editorial and production staff at Addison-Wesley, especially Kari Heen and Kim Ellwood who spent many hours on the phone with us. Our supervisors, Ann Rohovec, Bernie Piña, Scott Williams, and Kitty Berver, also have our heartfelt gratitude for allowing us the time and resources to pull this off and for their wonderful support. For technical support and help in producing a good quality book, we thank Chris Casey, Sandra Lee, Donna Alden, Judy Penna, Dru Sharp, Steve Finch, the reviewers of our manuscript, and a big thank you to Jon Stenerson at TCI Software Research, Inc. For help with making sure some of the applied activities were true to life and for ideas on some of the content of the book, we thank Dr. Eldon Steelman, Parker Bell, Dr. David Pengelley, Dr. Ross Staffeldt, and Dr. Lolina Alvarez. The instructors who helped us class-test these activities also receive our sincere appreciation: the college-track faculty in the Department of Mathematical Sciences at New Mexico State University, with special recognition to Sue Liefeld, Catherine Massey, Page Paneral, Emily Woods, and Jeannine Vigerust; the Developmental Mathematics faculty at Doña Ana Branch Community College, with special recognition to Bhagee Aiyer, Lucy Gurrola, Perry McDonell, Carolyn Moore, Susan Pfeiffer, Rene Sierra, Bobbie Thomas, and Frank Ureño. And, finally, we thank our students who inspired us to write this book and who have given us valuable input as collaborative learners. Many apologies to anyone we may have missed.

Book Reviewers

Lynn D. Darragh,
San Juan College

Susan MacLeod,
Greenfield Community College

James Douglas Robertson,
Bentley College

Sandra L. Spain,
Thomas Nelson Community College

Contents

Introduction: Managing Groups   1

    CREATING GROUPS FOR COLLABORATIVE LEARNING 2
    USING GROUPS FOR COLLABORATIVE LEARNING 4
    GRADING AND EVALUATION 8
    ISSUES AND ANSWERS 13

Key Activities for Initiating and
Monitoring Group Interactions   17

    Activity A: Think Teams 18
    Activity B: What Do Employers Want? 20
    Activity C: To the Student 22
    Activity D: Roles for Groups 24
    Activity E: How Are Things Going? 27
    Activity F: Using College to Reach My Goals 29
    Activity G: Group Review 32
    Activity H: Learning from Your Mistakes 35
    Cross-reference for Using Key Activities 37

Chapter 1: Basic Mathematics   39

    Activity 1-1:   Recycling 40
    Activity 1-2:   It's Party Time! 43
    Activity 1-3:   Stock Market 46
    Activity 1-4:   Fractions Step-by-Step 49
    Activity 1-5:   Do You Have Enough Money? 52
    Activity 1-6:   Sales Pitch 55
    Activity 1-7:   Making Comparisons 59
    Activity 1-8:   Thinking Along These Lines 63
    Activity 1-9:   News Report 66
    Activity 1-10:   Midterm Madness 69
    Activity 1-11:   The Survey 72
    Activity 1-12:   Using Measurement 75
    Activity 1-13:   A Moving Experience 78
    Activity 1-14:   Geometry Park 82
    Activity 1-15:   Follow the Signs 85
    Activity 1-16:   Thinking Aloud 88
    Activity 1-17:   "Algebragging" 91
    Activity 1-18:   Final Fling 94

Chapter 2: Introductory Algebra   99

    Activity 2-1:   Job Decision 100
    Activity 2-2:   Grams to Calories 102
    Activity 2-3:   Please Pass the Equation 106
    Activity 2-4:   Creating Applied Problems 109

Activity 2-5:   The Store Manager's Dilemma 111
Activity 2-6:   An Exponential Exploration 113
Activity 2-7:   Get That Factor Off My Back! 116
Activity 2-8:   How High Is That Rocket? 119
Activity 2-9:   Algebraic Fraction Puzzles 121
Activity 2-10: Graphing Charades 125
Activity 2-11: The Rent-A-Car Deal 127
Activity 2-12: Celsius vs Fahrenheit 130
Activity 2-13: Interpreting Results 133
Activity 2-14: Rent-A-Car II 136
Activity 2-15: The Search for the Perfect Square 138
Activity 2-16: Going a Round with Square Roots 141
Activity 2-17: The Maximum Playground 144
Activity 2-18: When Linear Meets Quadratic 147

Cross-reference of Algebra Activities   149

Chapter 3: Intermediate Algebra   151

Activity 3-1:   A Translating Team 152
Activity 3-2:   OOOP! Order of Operations Game 154
Activity 3-3:   Working with Sets 158
Activity 3-4:   Building a Road 160
Activity 3-5:   Parallel and Perpendicular Explorations 162
Activity 3-6:   Orange Juice Demonstration 165
Activity 3-7:   Coffee on the Run! 167
Activity 3-8:   Building a Sunroom 170
Activity 3-9:   Making a Bid 172
Activity 3-10: Water Works! 174
Activity 3-11: Complex Numbers 176
Activity 3-12: Not That Sunroom Again! 179
Activity 3-13: Getting Rational About Inequalities 181
Activity 3-14: A Carnival of Conics—Parabolas 185
Activity 3-15: A Carnival of Conics—Circles and Ellipses 188
Activity 3-16: A Carnival of Conics—Circles, Ellipses, and Hyperbolas 191
Activity 3-17: Functioning with Spreadsheets 194
Activity 3-18: How Much Space Do We Need? 197

ANSWER KEYS   201

Collaborative Learning Manual

Introduction: Managing Groups

To assist instructors in the process of implementing the collaborative learning activities contained in this book, the following information is provided in this introduction:

- Creating Groups for Collaborative Learning

    - Techniques for Forming Groups
    - Physical Arrangement of the Classroom

- Using Groups for Collaborative Learning

    - Roles to Facilitate Group Interactions
    - Getting Groups Started and Keeping Them Working
    - Student Responses to Group Work
    - Assessing Group Progress During an Activity
    - Closure After an Activity Is Completed

- Grading and Evaluation

    - Choosing Group, Individual, or Combined Grades
    - Participation Grade, Assigning Points, and Resubmitting Work
    - Evaluating Group Interactions
    - Assessing Individual Accountability
    - Incorporating the Collaborative Grade into the Course Grade

- Issues and Answers

- Key Activities for Initiating and Monitoring Group Interactions

    Activity A: Think Teams
    Activity B: What Do Employers Want?
    Activity C: To the Student
    Activity D: Roles for Groups
    Activity E: How Are Things Going?
    Activity F: Using College to Reach My Goals
    Activity G: Group Review
    Activity H: Learning from Your Mistakes

CREATING GROUPS FOR COLLABORATIVE LEARNING

Creating groups or teams of students to work together as problem-solvers adds variety and intellectual challenge to developmental mathematics courses. The process begins when student groups are formed and encouraged to work together in learning teams.

Techniques for Forming Groups

Several techniques can be used to create groups or teams of students to promote collaborative and cooperative learning in mathematics classrooms. Selection of a particular technique depends on the type of assignment the groups will be asked to complete, the size and "personality" of the class, the personality of the instructor, and the time available for establishing and implementing the group activity. Each activity contained in this book will refer to the grouping techniques listed next and will recommend the approach that is most appropriate for the assignment.

Proximity Pairing

This is the quickest and easiest method of initiating group activity. Asking students to work with the person next to them creates instant pairing that is useful for short-term assignments such as participating in get-acquainted exercises, comparing homework answers, or deciding on problem-solving strategies. Slightly larger groups can be formed by indicating specific rows or areas of the classroom. Minimal physical disruption to the classroom results from this method, since the students generally do not have to move desks or chairs. One disadvantage to this approach is that it does not take into account the specific needs and strengths of each student. It also somewhat limits the exchange of new information, since the students may be sitting next to someone they already know.

Number Off or Count Off

This tried-and-true technique has the advantage of working successfully with classes of any size. Students count off, out loud from 1 to n, where n is determined by dividing the number of students in the class by the number of students desired in each group. For example, if there are 31 students in the class and an activity requires about 4 people in each group, then the students should count off from 1 to 7, starting with the first student and continuing until each student has a number. This will create 4 groups containing 4 people each and 3 groups containing 5 people each. The technique does not require extensive preparation on the part of the instructor and it invigorates the classroom by requiring students to get up and move around to form the groups. This allows students to meet new classmates and even to see the classroom from a different perspective. This method is useful for short-term or limited-length activities where students can exchange ideas or compare assignment results. However, this method does not permit the instructor to address specific student needs or to consider the individual personalities of the students.

Structured Groups or Teams

Using structured grouping requires advance preparation by the instructor, as well as a good working knowledge of the *ability* and *personality* of each student in the class. Therefore this method is best applied after two or three weeks of class, when the instructor has had an opportunity to become acquainted with the students.

The advantage of structured groups is the establishment of balanced, productive groups capable of exchanging ideas and solving in-depth problems. Structured groups work well for long-term assignments requiring the students to meet and work together over several class periods. In that case, the group size should be kept small (no more than four students in a group) so that students can more easily mesh their schedules for meeting outside of class. Once structured groups are formed, the students may remain in them for the entire length of the course. Or, the instructor may choose to restructure the groups at some convenient point, such as after midterm or after the completion of a particular activity. Two common methods for structuring groups are ability mixing and ability matching.

- Ability Mixing

  Many group activities lend themselves to groups created by mixing the problem-solving and working speeds of the participants. A student who grasps concepts quickly can work with other students who may need help or more time to understand the problems. If possible, each group should be structured to contain students covering the full range of student achievement. To use typical grading letters, each group should have an "A" student, a "B" student, a "C" student, and so on. Groups should also be balanced in such a way as to allow quiet students to interact successfully with their more outspoken classmates. This balance of personalities is particularly important in developmental classes, where students may lack confidence in expressing their ideas.

  Ability mixing benefits students in several ways. The students who work faster can gain self-esteem by helping others. The slower students are encouraged to seek information from their peers rather than only from the instructor. All of the students have an opportunity to remember the story of the race between the tortoise and the hare. Slower is not always bad and faster is not always good! Some interesting lessons can be learned as students see the value of careful reading and cautious calculating, and learn the danger of skipping steps or doing sloppy work.

  The instructor will need to monitor these groups to ensure that all of the group members are involved and participating. The value of collaborative learning will be lost if the faster students do all of the work, or if the students with the higher grades feel that they must carry the entire group.

- Ability Matching

  Groups can be formed to allow students with the same ability level to learn together. If students are allowed to choose their own partners or to form their own groups, they often pick classmates who are "on the same wavelength" and work at the same speed. The instructor can structure the groups to build on this natural tendency.

  Ability matching can yield greater fairness in grading, since it brings together students who are accustomed to receiving similar grades. Students may also feel more comfortable in this type of group. Slower students may be reassured by the chance to tackle problems together step-by-step without rushing, while the faster students may enjoy making intuitive leaps. Students who finish their assignments early can assist their classmates by functioning as peer tutors or by writing problem solutions on the board.

Matched ability groups containing students with lower abilities will need more instructor assistance in class, as well as more tutoring outside of class.

Physical Arrangement of the Classroom

Group activities are most successful when students can work and sit in small circles or similar configurations to allow for good eye contact between the participants. Classrooms with movable chairs, desks, or tables permit the greatest flexibility. However, it is still possible to use group activities even in classrooms with fixed chairs that are bolted to the floor.

Flexible Classrooms

Allow time for students to move chairs, desks, and personal belongings each time a group activity is started. If an entire class period or larger block of time will be devoted to a group activity, remind students to sit with their group members each time the class meets. (*Note:* If multiple instructors use the same classroom, it is considered professional courtesy to return the classroom furniture to its original arrangement before the next class period. The students can do this before they leave the room. It is also possible to talk to the other instructors and explain the group activity and the need for some movement of furniture. The other instructors may be willing to use the same arrangement. This may start an exchange of ideas between instructors.)

Rigid Classrooms

It is tempting to rip the chairs out of the floor in this type of classroom! Since that is not usually practical, the best approach is to use Proximity Pairing. Up to four students can work together if two of them sit above and behind the other two students. This means some squirming and straining on the part of the participants, but it is possible. Also, look carefully at the aisle space in the classroom. Are there stairs or walkways where students can gather without blocking exits or violating fire codes? Don't be afraid to risk movement and disruption in the rigid order of the room. *The learning process is not limited to rows!* When people are actively involved in the collaborative learning process, they don't care whether they are sitting on the floor, standing in the aisle, or hunched over a desk.

USING GROUPS FOR COLLABORATIVE LEARNING

After the student groups have been formed, it is important to help them work together effectively. Collaborative learning does not occur just because groups have been formed. It takes time and instructor involvement to develop a cooperative learning atmosphere and to build cohesive, interactive groups. Topics to consider include the following:

- Roles to facilitate group interactions

- Getting groups started and keeping them working

- Student responses to group work

- Assessing group progress during an activity

- Closure after an activity is completed.

Roles to Facilitate Group Interactions

Many studies show that groups of three to four students are ideal, especially for longer activities that involve time outside of class. The instructor's task is to make sure the personalities of the students in each group mesh into a productive group interaction. At first, the group interaction may need to be controlled. Assigning group members one of four roles will help them get a feel for how to work together effectively. This may not be necessary for short, in-class assignments, but it works well for long-term activities. Suggested roles are:

- Moderator: Keeps the group on task and helps the group solve the activity by asking appropriate questions and encouraging everyone to participate.

- Quality Manager: Makes sure the work and the finished activity is the best the group can produce.

- Recorder: Keeps track of ideas and solutions during the group interaction.

- Messenger: Interfaces with the instructor by asking questions, or with the rest of the class by reporting results.

The students can switch roles or combine roles, if necessary. Activity D: *Roles for Groups*, can be used to help students get started in their assigned roles. Eventually, the groups should fall into a natural rhythm of working together with the four roles represented as appropriate, but with flexibility.

Getting Groups Started and Keeping Them Working

It is vital that you explain the assignment to the whole class, BEFORE forming them into groups. Make sure the entire class understands the activity before they move into groups. This allows the groups to start working quickly and eliminates a great deal of confusion.

Once the groups are formed, it is the instructor's task to keep moving about the classroom. It is important to facilitate the group process and to provide guidance. The instructor should:

- Observe progress in an unobtrusive way.

- Ask leading questions if a group is stuck for some time on one part of the activity.

- Keep students on-task and working effectively.

- Respond to student questions.

The instructor will also need to help the students learn how to work as a team. One way that students may naturally decide to tackle a lengthy problem is to assign portions to each group member, work on their portion individually, slap it all together, and call it done. This is the least helpful use of groups, since the point of forming groups is to work together. Activity C: *To the Student* and Activity D: *Roles for Groups* can be used to help the students understand and begin the group process. The students should:

1. Make a group plan for finishing the activity. Students will need to decide when each part should be done and in what order. If portions of the assignment are due at intervals, break these portions down into manageable tasks.

2. Work on each part, either together or individually. Students will need to decide together the best approach for solving each part or individually look over each part and compare methods to determine which one may be the best approach.

3. Synthesize the results of each part. Students will need to review and reread the activity as a group. Then, they should put the pieces together in a coherent way, making sure that the transition from one part to the next is smooth and logical.

These strategies may not be required in detail for a short, in-class activity, but some elements still apply. The point is to help the students manage their work in a productive and fair manner so they get the work done on time and so no one student is generating more effort than the others. Also, these strategies can help to incorporate peer teaching so the stronger students are helping the weaker ones.

Student Responses to Group Work

The single most important factor in the student response to collaborative learning is the attitude of the instructor! An instructor who communicates enthusiasm for group work, who respects developmental students as experienced, talented adults, and who prepares lessons with care will model for students the exact kind of behavior needed to achieve success. Possible responses of students to group work include:

The Outgoing, Talkative Student
This student will be happy to be part of a group effort and may naturally assume a leadership role in the group. The instructor will need to monitor this student to ensure that other, quieter students have an opportunity to speak during group sessions. It may also be necessary to remind the talkative student to stay on task so the group work can be accomplished.

The Quiet, Shy Student
This student will often blossom in small-group settings. He or she says, "I'm afraid to ask questions in class, but I don't mind talking in a small group." The shy student also benefits from the socialization that takes place during group activities as students get to know one another.

The Reluctant or Angry Student

Developmental students may bring personal problems with them into the classroom. Some may also feel frustrated by trying to learn math or embarrassed to be in a "remedial" math class. Introducing a new technique into the class or asking a student to pair up with other students can seem threatening to someone already overwhelmed with conflicting feelings. Refer to the section on Issues and Answers to get some ideas on how to assist this student.

The Procrastinator or Absent Student

Some students will come to group sessions unprepared or have irregular attendance during group activities. Even if they have valid reasons for their behavior, their unreliable participation can cause frustration for the group and a lack of productivity. Refer to the section on Issues and Answers for suggestions on how to handle this situation.

The Entertainer or the Worrier

Occasionally a student will want to spend time entertaining the group with stories or worrying aloud about personal problems. In either case, instructor intervention may be needed to help keep the group on task.

Assessing Group Progress During an Activity

As the groups become involved in completing an activity, it is necessary to monitor the progress of each group. Knowing which groups are on target and which are lost or falling behind will give the instructor an opportunity to spend time with the groups that need assistance. The instructor can:

- Ask for oral or written updates from each group. This technique is especially useful when groups are working on a long-term assignment.

- Ask the group recorder and/or messenger to provide a copy of the group calculations and problem-solving strategies at various points during the activity. This will help to detect any problem areas before the activity is completed.

- Watch for students who appear to be experiencing exceptional difficulty in understanding what is happening in the group. Be ready to assist individual students who might get "bogged down" in the mathematics.

Closure After an Activity Is Completed

While students are working on an activity, the instructor's role is that of a facilitator moving from group to group, but that role changes at the conclusion of the activity. At that point, the instructor will need to alter the classroom environment and guide the students from the interaction of a small group to the relationship of the large class as a whole. This includes eliminating the background noise that occurs during a group activity by returning to relative quiet. Such closure is necessary when the instructor intends to move from a collaborative process to a lecture process. It is also necessary to allow the class as a whole to discuss the results of the activity they have just completed. The instructor can conclude an activity and call the class to order in several ways:

- State a specific time when the activity must be completed. Make an announcement that indicates the amount of time left to complete the activity so the groups understand the approaching deadline.

- Call the class to attention. Tell the students to stop collaborating, or talking among themselves, and start listening to the instructor again. Ask the students to turn their chairs to face the front of the classroom or any other intended focal point, such as the board or overhead screen.

- Remind the class to follow the usual rules of courtesy as they listen to the next speaker, whether that person is the instructor or a fellow classmate.

Closure after an activity is completed also involves helping the students to realize what they have learned or accomplished by doing the group activity. This reinforcement of mathematical concepts and skills is a valuable part of the collaborative process. There are several methods an instructor can use to provide this reinforcement:

- Ask each group to share ideas and report results orally to the class.

- Lead a class discussion covering the same topic as the group activity. Ask questions that will assist the students in thinking about the concepts covered in the activity.

- Ask each student or each group for a brief, written summary of the results of the group effort. Exchange these statements so each group can read the results of other groups.

- Relate the group activity to the next topic of study in the course, to issues currently in the news, to certain chapters in the textbook, or to questions to be found on the next test. Students benefit from understanding the specific relevance of the group activity to their own lives.

- Allow the students to use the activity as a reference during a quiz or test on the same material.

GRADING AND EVALUATION

In a collaborative learning environment, the instructor can choose from a variety of grading and evaluation methods. Selection of a specific method depends on the type of collaborative activity the students will be asked to complete and the evaluation preferences of the instructor. Each activity in this book will refer to the types of grades and grading options listed in this section. Topics to consider include the following:

- Choosing group, individual, or combined grades

- Participation grade, assigning points, and resubmitting work

- Evaluating group interactions

- Assessing individual accountability

- Incorporating the collaborative grade into the course grade.

Choosing Group, Individual, or Combined Grades

Different types of grades may be given for group assignments. Each type has advantages and disadvantages.

Group Grade

A group grade means that all the students in one group receive the same grade for a specific group activity. A group grade may be given when one group paper is turned in for the assignment or when the group's work is graded as a whole even though each student in the group may have done a different part of the assignment.

- **Advantages**: Students are required to work together to complete the assignment. This also sets up a situation similar to most work environments. Using a group grade generates fewer documents for the instructor to grade.

- **Disadvantages**: Students may feel the grade does not accurately represent their personal effort. It is possible for a student to get a grade he or she has not earned. Also, the instructor may not be able to assess each individual's understanding of the material.

Individual Grade

Each student's work is graded separately. An individual grade may be given when each student has done his or her own assignment or a portion of the assignment while working in a group.

- **Advantages**: Students feel the grade more accurately represents their effort. This approach also has some built-in accountability. It works well for open-ended questions, where different answers are possible depending upon each student's situation or preference.

- **Disadvantages**: This approach does not provide an incentive for students to work together. It may not reflect how well the group worked together.

Combined Grade

This grade is based on both individual work and group work. A combined grade may be given in the following ways: (1) Students receive an individual grade for their portion of the assignment, averaged with a group grade for part or all of the assignment; (2) students receive an individual grade for their completed individual assignment averaged with the group's grade average; or (3) students receive an individual grade determined by how much each student contributed to the group effort, averaged with the group grade for the assignment.

- **Advantages**: This approach encourages group activity while rewarding individual effort. It has some built-in accountability and sets up a situation similar to most work environments.

- **Disadvantages**: This approach is complicated to explain to students. Also, calculating a combined grade will take more instructor time.

Participation Grade, Assigning Points, and Resubmitting Work

Instructors can choose different grading options for collaborative activities. Each activity in this book will refer to the grading options that follow.

Participation Grade

This grade is based on whether the student was actively involved in the activity, rather than on how well he or she did the activity. A participation grade may be given for an assignment that has no right or wrong answers, an assignment that replaces lecture on a topic, an assignment that is structured so that all groups find the correct solution, an in-class activity, or an overall assessment of the student's participation in the class work. A participation grade can take on many forms and be made up of many different grades and evaluations.

- The instructor may record participation or group interactions daily in a record book.

- The students may keep a notebook that holds all in-class activities and is graded periodically during the semester.

- The students may hand in a paper for each in-class or in-lab activity that will be incorporated in the participation grade.

Using a participation grade moves the focus from the grade to active learning. It rewards the students' participation without the penalty of getting a "bad" grade.

Assigning Points

Before assigning points, determine the goals of the assignment. Some possible goals are encouraging group activity, understanding a concept, demonstrating competence in a mathematical skill, and/or reviewing or connecting different mathematical concepts. The instructor will want to make sure that the selected point distribution will reflect the goals of the activity. There may also be nonmathematical aspects to the activity that should be included in the grade, including writing, art work, report preparation, or oral presentation. Most activities will use one or more of the following categories for distributing points:

- Mathematical calculations

- Report quality

- Clarity of explanations

- Creativity and extra effort.

For most developmental students, it is important to have points for neatness, style, and clarity to encourage them to submit college-quality papers. Rewarding creativity and extra effort will encourage students who often have difficulty in mathematics courses.

Resubmitting Work

Developmental mathematics courses mix students with widely varying abilities. Activities that may seem exciting and challenging to some students may discourage others. There are various reasons for allowing students to resubmit their work.

- Students who know they will be able to resubmit work will be more willing to tackle problems that at first may seem overwhelming.

- Allowing for resubmission keeps all students engaged and raises the standards for the class. An instructor may require work below a C to be resubmitted.

- Resubmitting solutions is more like the work world. Employers will ask their employees to redo a task until it is done properly or up to standards.

The instructor will need to determine how to handle the resubmission grade. Here are some options:

- Average the new and the earlier grades.

- Use a weighted average of the two grades.

- Totally replace the earlier grade with the new grade.

Evaluating Group Interactions

Besides grading the group activities, it is important to evaluate the group interactions. Since employers are becoming vocal about the need for employees to work well with others, it is important to monitor how well the groups are working together. Some suggestions on how to evaluate the group process follow:

- Use your own observations during in-class group work and assign a participation grade for that activity or day.

- Use individual questionnaires about how the students would evaluate their group interactions (see Activity E: *How Are Things Going?*).

- Talk to students individually during office hours.

- Evaluate the quality of work turned in by the group.

Assessing Individual Accountability

Many of the activities in this book have built-in accountability. Individual accountability is possible for all activities. To assess the work of individuals and to ensure that all students are working, consider these suggestions:

- Assign each group member a role as described in Activity D: *Roles for Groups*. If each student has an assigned role, he or she will be actively involved in completing the activity.

- Give a quiz on the basic points of the activity, and then adjust the activity grade based on each individual's quiz performance. This will give the instructor feedback for the entire class.

- Have each student estimate the percent of effort he or she and the other group members contributed to the completion of the activity. This can be in the form of a private, one-paragraph communication between the instructor and each student that is turned in with the completed activity. Give a combined grade to each student based on these estimates. If a student consistently contributes very little to the group effort, it is important for the instructor to talk with that student and, possibly, his or her group. If there is major disagreement among group members as to the amount of effort each contributed, the instructor may wish to meet with the group and find out whether they are getting along.

- Conduct student interviews during office hours or at another appointed time. Pick one student at random from each group to verbally explain the group's solution. This is usually an effective way to find out whether that particular student understands what is happening. Student interviews can also be used in conjunction with the preceding method to assess the understanding of a student who has contributed little to the group effort.

Incorporating the Collaborative Grade into the Course Grade

In developmental mathematics courses, group activities may form a small part (5% to 25%) of each student's course grade. The following suggestions for incorporating group activity grades include ideas used by several developmental mathematics instructors:

- Create a *separate category* for group activities, with a weight of 10% to 20% for this category.

- Include group activities as *part* of an existing category of grade. The most commonly used categories are homework and quizzes. Longer, out-of-class activities could be considered as a take-home exam. Average the group activity with all other activities in that category.

- Use group activities as *part of the participation grade,* which is usually 10% or less of the course grade.

- Count points from group activities as *extra credit*. In this case, each activity may be worth 5 to 20 points of extra credit. A maximum is usually established for the number of points each student can accumulate. The maximum generally constitutes 5% to 10% of the course grade.

Most instructors will find that they use more than one of these methods as different activities fit into different forms of grading. Some short, in-class activities may fit into a participation grade, while longer, out-of-class activities may be put into their own category.

ISSUES AND ANSWERS

- *Where do I find the time to use groups and still cover all of the material?*

    1. Rethink time. We have been conditioned to think about time in terms of the instructor and the material we must cover. Think instead of the time students spend learning. Is more student learning happening during lecture time or group time?

    2. Start slowly. You may want to find one or two areas in which you feel comfortable using groups, such as reviewing for exams or evaluating exams when they are returned. Look through the activities in this manual and pick some that particularly appeal to you.

    3. Lecture less but cover more. Trim your lecture down to the basics, and then make a mental or written list of items not in your lecture that you want to be sure the students understand. Watch for the students to discover these during group activities and class discussions. If this discovery doesn't happen on its own, then help it along by asking appropriate questions during the group activities or class discussions.

    4. Use group activities to replace lecture time. Allow students an opportunity to learn from one another and to solve problems in teams.

- *I'm not a group person!*

    1. Recognize that change comes slowly. Try one or two group activities that really excite you. Learn more about collaborative learning before you change your teaching style. Observe classes where group activities are used.

    2. Cooperate. Try team teaching with an instructor who loves groups.

    3. Learn from student suggestions. Students often have good ideas about how groups can work based on their experiences with group activities in other classes.

    4. Be kind to yourself. If groups do not work for you, do not use them. Avoid communicating a negative attitude about groups to your students.

- *I learned math in a lecture format, and it worked for me.*

    1. We all have different learning styles: auditory, tactile, and visual. Group activities allow students to talk, touch, and learn in new ways. This is particularly important for students in developmental classes.

    2. Groups allow some students to succeed who used to fail. Quiet students often blossom in small groups. They may participate by asking more questions than they would in a large class.

- *Developmental students are not ready for in-depth activities.*

    1. Try a new approach. Most developmental students have already had the lecture approach without fully learning the material. A new approach catches their attention.

2. Group activities may be closer to experiences outside the classroom. Many of the activities in this book relate mathematics to everyday life and give students a reason for learning a subject for which they have had no previous motivation.

3. Do it your way to fit your classes. As you assess each of your classes, you can decide how much material to cover before having students do a group activity on that subject.

4. Don't sell developmental students short! Developmental students are experienced, talented adults who bring many skills into the classroom.

- *What about the student who dominates the group?*

1. Talk about it. Have the class discuss problems that have arisen during the group work. Brainstorm as a class for solutions.

2. Take turns. The more you use groups, the more you will become familiar with your students. Assign domineering students to more passive rolls in a group (i.e., they may record what happens but not report).

3. Talk with the student. You may want to talk to this individual outside of class and help him or her find a more constructive way to participate in the group.

- *What about the student who is reluctant to work with other people?*

1. Explain the benefits of doing group work and model a positive approach. Activities A: *Think Teams* and B: *What Do Employers Want?* help the students become aware of how working together is a skill most jobs and careers require. Use your own experiences to show how diverse people can work effectively together.

2. Talk to the student privately. Are there other pressures in his or her life? Perhaps the student may benefit from the college support services in personal counseling, financial aid, and math tutoring.

3. This may not be the time for this student to work in groups. Allow the student to work individually on the same assignment or a different assignment.

- *Collaboration is not possible if students have irregular attendance or if they procrastinate and come to class unprepared.*

1. Create a "late group" and group the procrastinators or absent students together. Make it clear that there will be a penalty, such as 10 points deducted from their score, for any work that is submitted late.

2. Structure grading so one student in a group can fail but others can still succeed. Even in a cooperative work environment, the individual still must assume responsibility for his or her performance.

3. Call absent students and talk to them about the assignment. Encourage group members to exchange telephone numbers so they can keep in touch outside of class. If the student does not have a telephone, ask for a phone number where a message can be left.

4. Use positive peer pressure. Student group members are in the best position to make it clear that a true team effort requires everyone to participate.

- *Why aren't you teaching the class?*
Students may feel that the instructor isn't doing his or her job if the students are working in groups.

    1. Did you use those activities in this book that introduce group work to the class? Using Activity A: *Think Teams* or Activity B: *What Do Employers Want?* will help students understand some of the reasons for learning in groups.

    2. Give an explanation that focuses the class on "learning" instead of "teaching." This approach may allow students to see that learning increases when groups are involved. Emphasize that the goal is learning, and learning can come from people other than the instructor. By decreasing their reliance on the instructor, students are given the opportunity to increase their capacity to learn beyond the classroom.

    3. Mention that groups are not less work for the instructor. Many students will already realize this on their own when assignments have been returned with comments as well as grades. Let them know what you do to prepare for class.

- *Help! This group isn't talking or working!*

    1. Talk to the group about the reason. Is there one person who is dominating and consequently the others aren't working? Are they all reluctant to work with others? Are they all incredibly shy and no one has found a way to break the ice? What suggestions do they have for improving the group?

    2. Let the group get a divorce. If a long-term group is running into unresolvable difficulties, you may want to allow each group member to join another group.

    3. Let students evaluate the group process. Use Activity E: *How Are Things Going?* as a way to measure student response to group work. The results can be the basis for a class discussion.

- *Help! The students are trying to take over my class!*

    1. Group work empowers students. They may even begin to tell you how to teach. Be prepared for this sometime after midterm. You may need to reassert control.

    2. Define roles clearly. Let students know what their roles are in groups and what your role is as the instructor.

    3. Talk about it. When the class has a discussion about groups, let them know your view of the situation.

Key Activities for Initiating and Monitoring Group Interactions

The activities in this section are designed for any course in which an instructor wants to use collaborative learning. The topic for each activity is listed so the instructor may more readily decide which ones he or she would like to use.

Activity A: Think Teams   Why Work in Groups?

Activity B: What Do Employers Want?   Qualities Employers Value

Activity C: To the Student   Steps for Working in Groups

Activity D: Roles for Groups   Steps and Roles for Working in Groups

Activity E: How Are Things Going?   Evaluation

Activity F: Using College to Reach My Goals   Setting Goals

Activity G: Group Review   Mathematical Exam Preparation

Activity H: Learning from Your Mistakes   Evaluating Math Mistakes

NOTES TO THE INSTRUCTOR

Summary: The question "Why work in groups?" is answered through class participation. Students discover the key qualities employers value in their employees. Those qualities are then related to activities, such as group work, in the math classroom.

Skills Required: None
Skills Used: Reason for Working in Groups

Grouping: The entire class (This exercise is used *before* group activities begin.)
Materials Needed and Preparation: Copies of Activity A, one for each student
Student Time: [**In Class**] 10 to 15 minutes; [**Out of Class**] none

Teaching Tips: This brief activity is used to set the stage for successful group work and can be used immediately before dividing the class into groups. It is particularly useful before starting long-term group assignments that span several class periods. Begin by asking students to say aloud why they decided to attend college. The most common reason is usually to prepare for a job or a career. Using the printed activity sheet, ask students to list, in order, the three most important qualities they think employers want in their employees. Ask the students to share their number one quality, then write on the board or an overhead the quality employers chose. Do this for qualities two and three. Then direct a brief discussion on how this math class can help to prepare the students to improve or achieve each of those qualities. Students should fill out the activity sheet and keep it as a reminder of one of the reasons for working and learning in teams—to achieve success in the workplace. Other rewards of collaboration may also be mentioned, such as making friends, practicing English, overcoming anxiety, sharing ideas, and learning it is okay to make mistakes. OPTIONAL: Ask students to list on the bottom of the activity sheet all the ways they will interact with other people in their chosen career fields.

Grading Tips: This activity is not graded.

Comments: Adult students usually have a direct interest in being employable and successful in the workplace. This activity helps students think about the real-world applications of learning to solve problems and working together in group settings. You may want to share your own career experiences with committee work, team projects, and the like, especially as it pertains to learning to work successfully with all kinds of people in all kinds of situations.
Connections: Activity B: *What Do Employers Want?* and Activity C: *To the Student* or Activity D: *Roles for Groups*

One reason for attending college is to prepare for a job or career. Students expect college classes to provide instruction and information to help them prepare for their chosen career fields. However, many important skills and qualities that are required for success in the workplace may not be covered directly in the course content.

Employers are frequently asked to describe the qualities they consider most important in a good employee. The top three qualities employers want in their workers are as follows (make your best guess):

1. _____

2. _____

3. _____

This math class will provide you with opportunities to develop skills and build strengths in all three of those areas in the following ways:

1. _____

2. _____

3. _____

Think Teams allow students to work and learn together both in and out of class. Teamwork helps people share ideas, solve problems, and reach goals. If you have decided on a career, think of all the different ways you will be interacting with other people as you work!

NOTES TO THE INSTRUCTOR

Summary: Students are given a list of qualities employers value in their employees. The students rank the qualities and compare their choices with the results of an actual employer survey. The results can be used to explain the use of collaborative learning as an instructional technique.

Skills Required: None

Skills Used: Qualities Employers Value, Reason for Working in Groups; OPTIONAL: Personal Assessment for Employability

Grouping: Count Off or Proximity Pairing

Materials Needed and Preparation: Copies of Activity B, one for each student

Student Time: [**In Class**] 5 minutes on Day 1, 10 to 15 minutes on Day 2; [**Out of Class**] short homework assignment

Teaching Tips: Distribute Activity B on Day 1 and ask the students to rank the qualities as a homework assignment. On Day 2, ask students to form small groups before they know the actual ranking given by the employers. The students can discuss their choices and explain their reasons for ranking the qualities as they did. After a few minutes of small-group discussion, explain the ranking that the employers gave and discuss any differences with student rankings. Emphasize the use of group activities in your math class as a way for the students to practice working with other people and to develop leadership potential as they prepare for their careers. OPTIONAL: To personalize this exercise for students, continue the assignment with these new instructions: "Please assess yourself in each of the above qualities on a scale from 1 to 5 ('1' indicating 'I am very skilled' and '5' indicating 'I am not skilled')." This scale can be written in the left margin of the activity sheet. The students can then practice goal setting, which employers also value, by choosing one or two of these qualities to work on during the rest of the course. They can write a brief statement outlining their goals on the bottom of the activity sheet.

Grading Tips: This activity is not graded. If the optional personal assessment is used, the goal statements can be given extra credit or participation points.

Comments: Students are usually very interested in knowing what qualities employers look for in an employee. This activity provides an excellent opportunity to help students identify some of those qualities. The information is taken from the Lindquist-Endicott Report, by Victor R. Lindquist, Northwestern University Placement Center, Evanston, Illinois. This activity helps students understand that successful employment is dependent on more than grades and technical skills. It also helps students see that working in groups in the classroom is a useful preparation for meeting employer expectations.

Connections: Activity A: *Think Teams*; Activity C: *To the Student*; Activity D: *Roles for Groups*

Groups   Qualities Employers Value

(Source: The Lindquist-Endicott Report, by Victor R. Lindquist, Northwestern University Placement Center, Evanston, Illinois)

Listed below are some qualities employers typically mention that they are looking for in employees. Imagine yourself as a personnel director who wants to hire a college graduate for his or her firm. What are the qualities you will be looking for in the applicants? Rank the qualities below in order using "1" to indicate "most important" and "13" to indicate "least important."

	Rank	Qualities
a.	_____	Ability to get along with people, self-confidence, and leadership potential
b.	_____	Academic achievement, grades
c.	_____	Well thought-out personal and career goals
d.	_____	Interest in the type of work my company performs
e.	_____	Personal initiative, enthusiasm, drive
f.	_____	Ability to express him/herself orally and in writing
g.	_____	Realism of salary demands, willingness to work up from the bottom
h.	_____	Personal appearance
i.	_____	Involvement in extracurricular activities
j.	_____	Awareness of what my company is all about, has read literature about it
k.	_____	Willingness to travel or move
l.	_____	Concern for security and benefits versus interest in job
m.	_____	Preparation for position for which he/she is applying

NOTES TO THE INSTRUCTOR

Summary: Students are given a list of steps to help them get a successful start on working and learning in groups. Space is provided for recording the names of the group members.

Skills Required: None
Skills Used: Steps for Working in Groups

Grouping: Any technique (Activity C is given to the students *before* they form groups or concurrently with group work.)
Materials Needed and Preparation: Copies of Activity C, one for each student; OPTIONAL: The activity can be printed on the back of either Activity A: *Think Teams* or Activity B: *What Do Employers Want?*
Student Time: [**In Class**] 5 to 10 minutes; [**Out of Class**] none

Teaching Tips: This activity can be used concurrently with any of the group activities in this book. Distribute the group assignment along with Activity C before breaking the class into groups or teams. Allow time for the students to scan the steps in Activity C. Explain that they will use this list of steps to begin working in groups today. Form the class into groups using any grouping technique. Remind students to record the names of their group members and to follow the remaining steps in the list.
OPTIONAL: Use Activity A or Activity B to help students understand how group work is valued by employers. Then distribute Activity C, *To the Student* (or refer to it if it is printed on the back of Activity A or B). Explain the use of these steps for successful group work.

Grading Tips: This activity is not graded.

Comments: This activity is especially effective when used in conjunction with Activity A or Activity B. It helps students understand that the interpersonal skills they may develop by working on math group activities will also be useful in the workplace.
Connections: Activity A: *Think Teams;* Activity B: *What Do Employers Want?;* Activity D: *Roles for Groups*

If "two heads are better than one" to solve a problem, how about three heads, or four? Teamwork builds buildings, makes music, explores space, creates computers, performs surgery, and wins games. Go for it!

1. The first step in working with other people is to get acquainted. Introduce yourself to your group. Write the names of your group members here:_____

   _____

2. TALK about the assignment. What is it about? What is your group supposed to do? When is the activity due to be turned in?

3. Read the assignment again and ask questions. Is there something you don't understand? Ask! Can you answer questions for other group members?

4. Get started. Jump in and throw out ideas or make a stab at solving part of the problem. Compare notes and ideas with your teammates.

5. LISTEN to your teammates. Follow the usual rules of courtesy and respect. Even a "bad" idea can lead to a great way to solve a problem!

6. ENJOY the opportunity to meet new people, hear new ideas, and work together to solve problems.

NOTES TO THE INSTRUCTOR

Summary: Students are given a list of steps and roles to help them work and learn successfully in groups. Space is provided for recording the names of the group members as well as role assignments.

Skills Required: None
Skills Used: Steps for Working in Groups

Grouping: Any technique (Activity D is given to the students when they form groups and concurrently with group work.)
Materials Needed and Preparation: Copies of Activity D, one for each student
Student Time: [**In Class**] 5 to 10 minutes; [**Out of Class**] none

Teaching Tips: This activity is designed primarily to be used concurrently with any of the group activities that require an extended time for group interactions or more structured interactions. Distribute the group assignment along with Activity D before dividing the class into groups. Allow time for the students to read Activity D as well as the group assignment. Form the class into groups using any grouping technique. Ask students to record the names of their group members and role assignments and then to follow the remaining steps in the list while working on their assignment.

Grading Tips: This activity is not graded.

Comments: Use this activity in conjunction with Activity A or B. All of these activities help students understand that the interpersonal skills they may develop by working on math group activities will also be useful in the workplace. This activity, along with Activity C, helps students develop these interpersonal skills.
Connections: Activity A: *Think Teams;* Activity B: *What Do Employers Want?;* Activity C: *To the Student*

Working in groups for the long term is a good way to make friends and develop confidence. You will have a built-in study group and you may meet new people with whom you share interests. Below are some strategies to help your interactions be productive and fun!

1.   INTRODUCE yourself. Write the names of your group members here:_____

_____

2.   ASSIGN roles. Write the name of the assigned group member next to his or her respective role.

   • Moderator: Keeps the group on task and helps the group solve the activity by asking appropriate questions and encouraging everyone to participate.

   NAME:_____

   • Quality Manager: Makes sure the work and the finished activity is the best the group can produce.

   NAME:_____

   • Recorder: Keeps track of ideas and solutions during the group interaction.

   NAME:_____

   • Messenger: Interfaces with the instructor by asking questions, or with the rest of the class by reporting results.

   NAME:_____

3.   TALK about the assignment. What is it about? What is your group supposed to do? When is the activity due?

4.   READ the assignment again and ask questions. Is there something in the assignment you don't understand? Ask! Can you answer questions for other group members? Does the messenger need to get more information from the instructor? Can different people on the team find out information that everyone can use to solve a problem?

5.   DIVIDE AND CONQUER!

   • Make a group plan for finishing the activity. Decide when each part should be done and in what order. If portions of the assignment are due at intervals, break these portions down into manageable tasks. Decide whether this activity needs to be divided into smaller parts.

- Work on each part, either together or individually. Decide together the best approach for solving each part or individually look over each part and compare methods to determine which one may be the best approach. WARNING: You are responsible for understanding the whole assignment, so resist the temptation to let other group members do the work for you. That spoils the team effort.

- Synthesize the results of each part. Review and reread the activity as a group and determine whether you have completed the assigned task. Then put the pieces together in a coherent way, making sure that the transition from one part to the next is smooth and logical.

6. LISTEN to your teammates. Follow the usual rules of courtesy and respect. Even a "bad" idea can lead to a great way to solve a problem!

7. PLAN to meet again. Ask the instructor whether additional class time will be available for group meetings or whether you will need to meet outside of class. (If your group will be working together for several class periods or weeks, you can exchange telephone numbers and schedules or make plans to stay in contact outside of class.)

8. ENJOY the opportunity to meet new people, hear new ideas, and work together to solve problems.

Activity E: How Are Things Going?

NOTES TO THE INSTRUCTOR

Summary: Students are given an opportunity to evaluate group processes and to assess their own contributions to group activities.

Skills Required: Experience Working in Groups
Skills Used: Process Evaluation, Critical Thinking

Grouping: The entire class
Materials Needed and Preparation: Copies of Activity E, one for each student
Student Time: [**In Class**] 10 to 15 minutes; [**Out of Class**] none; OPTIONAL: individual homework assignment

Teaching Tips: Use this activity after several weeks of group work. Ask the students to fill out the activity in class, or to complete the activity as an individual homework assignment. After you have had an opportunity to collect the activity and read what the students have written, make two lists of their responses to questions 4 and 5 on the board. Discuss the negative items from question 4, and ask the students to think of ways to reduce any problems they have encountered while working in groups. Discuss the positive items from question 5 and add anything that you feel is missing from the list. You may want to discuss other topics as well. For example, it may be interesting to compare the responses to questions 8 through10 with the responses to questions 3 through 7 to see whether students who have negative feelings about group work think they have worked very hard or not at all hard on the group activities. Point out that asking someone else for help with a math problem (a common response to question 1) is an example of using a group process to solve a problem. If you have not already used Activity A: *Think Teams* or Activity B: *What Do Employers Want?* you may want to do so soon.
Possible additional topics for discussion:
1. Why do employers place such a high value on working well with other people?
2. How does problem solving in a group differ from problem solving alone?
3. How does learning in a group differ from learning alone?
4. What can be done to ensure that everyone makes a contribution to the group effort?

Grading Tips: This activity is not graded, although it may be used for a Participation Grade.

Comments: Although employers seek employees who work well with others, college classes usually provide few opportunities for students to practice working and interacting with one another. Helping students to develop these skills also means giving them a chance to evaluate their own progress. This activity allows students to examine their own contributions to the group effort, and to communicate their opinions to one another and to the instructor.
Connections: Activity A: *Think Teams;* Activity B: *What Do Employers Want?;* Activity F: *Using College to Reach My Goals*

Activity E: How Are Things Going?

Groups _____ Evaluation

DO NOT PUT YOUR NAME ON THIS PAPER. Write as much or as little as you need to finish each sentence. Please be honest!

1. If I get stuck on a math problem, or if I need help understanding something, what I usually do is

2. So far, I have been absent and missed this math class (circle one)

    more than 5 times   3–5 times   1 or 2 times   never

3. When I'm part of a group or team and we are working together on an activity, what I usually do is

4. One of the things I *do not like* about working in groups or teams is

5. One of the things I *really like* about working in groups or teams is

6. The difference between working alone or working with other people to solve math problems is

7. Working in groups with other people in school and/or on the job is

USE THE SCALE ON THE RIGHT to choose the best answer for the following questions:

	Not Hard			Very Hard
8. How hard have I worked to be part of the group effort to solve problems during class time?	1	2	3	4
9. How hard have I worked to be part of the group effort to complete activities outside of class?	1	2	3	4
10. How hard is it for me to do the math in the group activities we have done so far in this class?	1	2	3	4

PLEASE USE THE BACK OF THIS PAGE for any other comments you would like to make about learning and working in groups or teams.

28

NOTES TO THE INSTRUCTOR

Summary: Students respond to questions about setting goals and how they can develop successful techniques in college to reach those goals.

Skills Required: None
Skills Used: Goal Setting, Critical Thinking

Grouping: Any technique
Materials Needed and Preparation: Copies of Activity F, one for each student
Student Time: [**In Class**] 5 minutes on Day 1, 10 to 15 minutes on Day 2; [**Out of Class**] individual homework assignment

Teaching Tips: This activity can be used any time during the course to help students focus on using college to reach their goals. Introduce the activity on Day 1, and allow students to read the introduction and the directions. Let the students know whether their responses to the questions on page 2 will be used as group discussion topics, or whether you intend to use the entire activity as a personal communication between student and instructor. On Day 2, discuss the answers to page 2 in groups or with the entire class. It may be useful to discuss the difference between "a dream" and "a goal." It may also be helpful to substitute the concept of "hitting a target" for the idea of reaching a goal. Possible additional topics for discussion on Day 2 include the following:
1. What is the difference between long-term and short-term goals?
2. What does it mean to prioritize goals?
3. How can a long-term goal be broken down into smaller goals?
4. How can you stay on track as you work toward a goal?
5. How will this math class help you achieve your goals?

Grading Tips: This activity is not graded, although it may be used for a Participation Grade. Also, you may use the results of the activity to make general comments to the class, or you may write personal responses on the activity sheets before you return them to the students.

Comments: Although this activity can be used any time, it is particularly effective at midterm when students may demonstrate a lack of focus and instructors may feel a need for improved lines of communication. Encouraging the students to write about their goals also encourages them to examine their learning strategies within your math class. In addition, it can be an informative experience for the instructor to learn more about the plans and goals of these "developmental" students.

Activity F: Using College to Reach My Goals Name:_____

INTRODUCTION:

About this time in the semester, the stress level begins to increase. There may be changes and pressures in your personal life as well as in your academic life. This is a good time to remind yourself *why* you chose to attend college and what you need to do to achieve your goals.

DIRECTIONS: Please answer these questions carefully and honestly. The questions on page 1 will be used as a way to communicate your goals and ideas to your instructor. The questions on page 2 may form the basis for a discussion with the members of your group or team. Ask yourself the following questions:

1. What do I want to do with my life? What do I want to learn, to experience, or to accomplish?

2. How important is my college education in reaching these goals?

3. In the past, how successful have I been in using my time and talents to reach my goals?

4. How can I translate my goals into specific actions in my daily life? What do I need to do to reach my goals?

5. How will this math course help me achieve my goals?

6. What techniques that have worked well for me in the past can I continue to use to reach my goals in college? What are my strengths? (Be specific!)

7. What *new* techniques do I need to learn to help me reach my goals in college? What ideas can I borrow from other people? (Be specific!)

NOTES TO THE INSTRUCTOR

Summary: This activity is designed to help mathematics students analyze their deficiencies and help prepare for a mathematics exam. It starts in class with an individual math study-needs analysis. This activity then helps a group of students plan an out-of-class study session, record the session, and do an individual evaluation of the study session.

Skills Required: None
Skills Used: Analyzing Mathematics Study Skills Needs, Preparing for a Mathematics Exam

Grouping: Structured Groups
Materials Needed and Preparation: Copies of Activity G, one for each student
Student Time: [**In Class**] 20 minutes; [**Out of Class**] time to meet at least once before an exam

Teaching Tips: Use this activity before a major exam. Make sure that it is completed more than one day before the exam. It can also be used informally with a group of students who regularly study together. You may want to create groups by locale (those who live in the same area), by preferred language, or by available meeting times. Have students hand in the activity when they come to take the exam.

Grading Tips: 25 points for each part; or consider the activity as the first problem on the exam or an extra credit problem on the exam.

Comments: Developmental students sometimes need to learn how to assess their own study needs. Studying in groups may not come naturally to some students. This activity addresses both of these concerns.
Connections: Activity H: *Learning from Your Mistakes*
Spin-offs: If done early in the semester, study groups may continue to meet without help from the instructor.

Activity G: Group Review

This activity is designed to help you develop a more effective way of preparing for an exam. The first and last parts are done individually. The second and third parts are done in groups.

Part I: Personal Mathematics Study-Needs Analysis

You are approaching a mathematics exam, or it is approaching you. Take charge by answering the following questions with this upcoming exam in mind.

1. Which topics will be included in this exam?

2. On which of the above topics do I feel confident?

3. On which of the above topics do I need more work?

4. Pick one of the following to finish the statement: "I study best _____."

   a. alone   b. with others   c. with a tutor

5. What resources are available to me for studying (i.e., textbook, old quizzes, class notes, homework)? Which of these have helped me the most?

6. On which of the following skills do I need work:

   Lowering test anxiety   Checking for errors   Memorizing formulas

   Reading and understanding the problem   Careful calculation

   Specific areas of arithmetic (such as fractions)

Part II: Group Comparison

Take your answers from Part I to your assigned group and continue.

1. Compare your answers from Part I with those of your group members.

2. Make a study plan with your group members. Include the topics and skills on which you will all work. Decide which resources you will each bring to the study session.

3. Arrange a study session. Pick a place and time when the members of your group can meet. Record this information as well as the names and phone numbers of your group members.

    Place _____   **Time** _____

        Names   **Phone Nos.**

        _____   _____

        _____   _____

        _____   _____

        _____   _____

Part III: The Group Study Session

Fill out the following record of the group study session:

Who attended (list names here)?

Where did you meet?

How long did you study?

Which topics did you study?

Part IV: Personal Evaluation

Answer the evaluation questions honestly. The purpose here is to help you find a method of study that works for you.

1. Did you find the group study session helpful? Why or why not?

2. For which topics or skills did you find the session most helpful?

3. With which topics or skills do you feel you still need help?

4. Would you like to participate in another group study session? Why or why not?

NOTES TO THE INSTRUCTOR

Summary: Students are paired when a homework or exam is returned and given a mistake evaluation form to help them understand the types of problems they missed.

Skills Required: Reading a Form
Skills Used: Analyzing Mathematical Mistakes

Grouping: Structured Groups, Ability Mixing (2 students per group)
Materials Needed and Preparation: Copies of Activity H, one for each student; plan groups. *Note:* There is a blank column heading on the activity where you can add your own question.
Student Time: [**In Class**] 10 to 50 minutes; [**Out of Class**] none

Teaching Tips: This activity is given when a homework assignment or exam is returned. Before handing back the assignment, hand out the mistake evaluation form. Spend some time going through the directions and doing an example for the students. Students are to look at each problem they missed and pick one of the reasons given for that mistake. They do not need to rework the problem but may if they are given enough time. If their mistake seems to be different from any of those listed, they can write in their own description of the mistake. Instructors may want students to write their own explanation in place of the choices given. Do not accept "stupid mistake" as a reason.

Grading Tips: Individual Participation Grade

Comments: This activity can take the place of the instructor's going over the exam. It can take as long as you want, as some students will need plenty of time. Those who do not finish in class can finish at home. Some teachers have voiced a concern that students would not like to let someone else see their grade. The author has never had a student object to doing this activity. The students with the better grades are usually quite willing to help those with the lower grades. Students with lower grades get some needed one-on-one help and often are less discouraged as they find a pattern to their mistakes. Instructors should be prepared for students to find any discrepancy that may have occurred in grading.
Spin-offs: Some students paired this way have continued as peer tutors for the rest of the course.

Activity H: Learning from Your Mistakes

As you continue to study and learn math, you will continue to make mistakes. Understanding and learning from these mistakes is a key to succeeding in math. Mathematicians and math instructors make mistakes. The trick is to catch the mistakes before you hand in your work. This evaluation form will help you to identify the types of mistakes you are making.

Assignment:_____   Name:_____

Type of Error (choose one for each problem missed): A = Arithmetic,
C = Carelessness, **M** = Misunderstood, **D** = Didn't know how to do it

Problem Number	Type of Error		Did you attend class?	Did you do homework?	Did you study this?

What was your most common type of error?

What could you do to prevent or catch that error in the future?

Cross-reference for Using Key Activities

The following cross-reference explains when to use each of the previous key activities. Also listed are specific mathematical activities in this book that reference each key activity.

Key activities	Uses and cross-reference
Activity A: Think Teams	Use *before* any activity in chapters 1 through 3 of this book, for example, activities 1-3, 1-7, 1-11, 1-13, and 2-2.
Activity B: What Do Employers Want?	Use *before* any activity in chapters 1 through 3 of this book, for example, activities 1-3, 1-11, and 2-2.
Activity C: To the Student	Use *before or in conjunction* with any activity in chapters 1 through 3 of this book. Especially useful for activities 1-7, 1-8, 1-13, 3-1, and 3-11.
Activity D: Roles for Groups	Use *before or in conjunction* with any activity in chapters 1 through 3 of this book. Especially useful for activities 1-11, 2-2, 2-5, 2-8, 2-12, 3-7, 3-8, 3-10, and 3-12.
Activity E: How Are Things Going?	Use *after midterm* or after several weeks of group work, for example, after Activity 1-7.
Activity F: Using College to Reach My Goals	Use anytime during the course or *after midterm,* for example, after Activity 1-10.
Activity G: Group Review	Use *before* a major test or exam.
Activity H: Learning from Your Mistakes	Use *after* any test, quiz, or homework assignment has been graded and returned to the students, for example, after activities 1-16 and 2-9.

Chapter 1: Basic Mathematics

The activities in this chapter are designed for the course in Basic Mathematics. Some of them may be appropriate for other courses as well. The topic for each activity is listed so the instructor may more readily decide which ones he or she would like to use.

Activity 1-1:	Recycling	Whole Numbers
Activity 1-2:	It's Party Time!	Fractions
Activity 1-3:	Stock Market	Fractions
Activity 1-4:	Fractions Step-by-Step	Operations with Fractions
Activity 1-5:	Do You Have Enough Money?	Decimals
Activity 1-6:	Sales Pitch	Decimals
Activity 1-7:	Making Comparisons	Ratio and Proportion
Activity 1-8:	Thinking Along These Lines	Percents
Activity 1-9:	News Report	Percents
Activity 1-10:	Midterm Madness	Midterm Review
Activity 1-11:	The Survey	Descriptive Statistics
Activity 1-12:	Using Measurement	U.S. and Metric, Perimeter and Area
Activity 1-13:	A Moving Experience	Volume and Area
Activity 1-14:	Geometry Park	Geometry Review
Activity 1-15:	Follow the Signs	Rational Numbers
Activity 1-16:	Thinking Aloud	Algebra
Activity 1-17:	"Algebragging"	Algebra
Activity 1-18:	Final Fling	After Midterm Review

NOTES TO THE INSTRUCTOR

Summary: Whole numbers, used by "real world" companies to report recycling efforts, are used for practice with basic number operations. Students relate data to their own lives and share information with their class partners.

Skills Required: Whole Number Addition, Subtraction, Multiplication, Division
Skills Used: Whole Number Word Names, Rounding, Exponential Notation, Estimation

Grouping: Proximity Pairing
Materials Needed and Preparation: Copies of Activity 1-1, one for each student
Student Time: [**In Class**] 5 to 10 minutes on Day 1, 10 minutes on Day 2; [**Out of Class**] individual homework assignment

Teaching Tips: On Day 1, introduce the activity. This timely topic is often in the news. Explain procedures and refer students to the textbook pages that cover the same material. Remind students to show their work on page 2 and to use correct units in their answers, to receive full credit, since answers will vary. For question 5, "family" can mean household, roommates, etc. Note that questions 8 and 9 will be answered in the next class (Day 2). On Day 2, use Proximity Pairing to group students. Allow time to compare answers and to solve questions 8 and 9.

Grading Tips: Individual Grade; 5 points for each of the 20 answers (excluding partner's name); OPTIONAL: reduced points for missing or inappropriate units

Comments: This activity helps students learn that whole number math is useful in the world outside the classroom and important for everyday issues in our own lives. Because whole numbers are covered at the very beginning of basic mathematics, this initial activity uses several techniques to assist the students. (1) Key words are in bold type and the pages are formatted with lines and boxes for answers to carefully guide nervous students through this first assignment. (2) Proximity Pairing gently introduces the idea of cooperative learning and encourages students to begin interacting with their classmates. (3) Leading statements are provided before some questions to help students begin the problem-solving process.
Connections: Refer to this activity during geometry. How much landfill space (volume) would be needed for all of this garbage!
Spin-offs: Discussion topics such as, "Does this community and/or college have a recycling program?"; "How many students participate in recycling activities?"

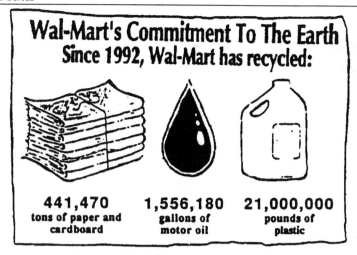

—Courtesy of Wal-Mart Stores, Inc., 1993. Reprinted with permission.

"Did you know that if Americans recycled all of our Sunday newspapers, we could save over 500,000 trees each week or 26,000,000 every year?" —Information from a 1993 Kellogg Company cereal box top, courtesy of Kellogg Company. Reprinted with permission.

Use the information on recycling shown above to answer questions 1, 2, and 3.

1. Use the space below to write the **word name** for each number. Wal-Mart has recycled:

    _____

        tons of paper and cardboard.

    _____

        gallons of motor oil.

    _____

        pounds of plastic.

2. Use **rounding** to complete the chart below. Wal-Mart has recycled approximately:

	To the nearest hundred	To the nearest ten-thousand	To the nearest hundred-thousand
tons of paper and cardboard			
gallons of motor oil			

3. **Exponential notation**, using 10 raised to a power, provides a short notation for large numbers. For example,

$$370,000 = 37 \cdot 10 \cdot 10 \cdot 10 \cdot 10 = 37 \cdot 10^4$$

    Use exponential notation to complete each statement.

    $21,000,000 = 21 \cdot 10 \cdot 10 \cdot 10 \cdot 10 \cdot 10 \cdot 10 = \underline{21 \cdot 10}$_____ pounds of plastic.

    $500,000 = 5 \cdot 10 \cdot 10 \cdot 10 \cdot 10 \cdot 10 = \underline{5 \cdot}$_____ trees each week.

    $26,000,000 = 26 \cdot 10 \cdot 10 \cdot 10 \cdot 10 \cdot 10 \cdot 10 = \underline{26 \cdot}$_____ trees every year.

Garbage Produced by 1 Person in the United States.

1,168 LBS.

Over
3 LBS.
Each
Day

96 LBS.
Each
Month

Each
Year

—Courtesy of Wal-Mart Stores, Inc.,
1993. Reprinted with permission.

Use the information on garbage production shown above to answer questions 4 through 9.
SHOW YOUR WORK, and use the correct units to label your answers.

4. In one year, if you recycled 279 pounds of garbage, how much of your garbage would be
   left that was not recycled?

   _____

5. How many people are in your family? How much garbage would be produced by your
   family

   each month?_____

   each year?_____

6. Use your own age (to the nearest year) to answer this question. Approximately how
   many pounds of garbage have you produced

   in your life?_____

7. There are 2,000 pounds in 1 ton. How many *tons* of garbage have you produced in your
   life? Use R to show any remainder.

   _____

8. Compare answers with your class partner. Partner's name_____
   Then work together to answer these questions. What is the total amount of garbage
   produced by your two families

   each month?_____

   each year?_____

9. **Estimate** the amount of garbage produced by the students in this math class

   each month._____

Activity 1-2: It's Party Time!

NOTES TO THE INSTRUCTOR

Summary: Actual recipes are used to involve students in manipulating fractions to plan for a party.

Skills Required: Fractions, Mixed Numerals
Skills Used: Multiplication and Division of Fractions and Mixed Numerals, Simplifying, Reasonableness of Answers

Grouping: Proximity Pairing
Materials Needed and Preparation: Copies of Activity 1-2, one for each student; OPTIONAL: measuring cups and spoons for use as visual aids
Student Time: [**In Class**] 10 minutes on Day 1, 10 to 15 minutes on Day 2; OPTIONAL: 30 minutes on one day; [**Out of Class**] individual homework assignment if done over 2 days

Teaching Tips: This activity may be done completely during one class period by allowing student partners to work together. Or, it can be introduced on Day 1 as a homework assignment to be completed with a partner in class on Day 2. Remind students of the meaning of mixed numerals and simplifying fractions. (Note the use of "SERVES ... PEOPLE" in each recipe as it relates to questions 2 and 4. To prevent leftovers in question 4, students should be multiplying by $4\frac{1}{2}$.)

Grading Tips: Individual Grade; 3 points for each of the 32 answers in questions 1 through 7; 4 points for question 8

Comments: Working with fractions can seem threatening and laborious to students, who usually prefer decimals. Using an application with a food and fun flavor can relate fractions to everyday living. Question 7 emphasizes the value of team effort and group work in problem solving, as well as stressing the need to follow written and oral directions. Teamwork helps to catch errors, such as overlooking the data on the number of people served by each recipe. Question 8 invites students to think about an issue that is critical in all of mathematics, but especially in working with fractions—how to decide whether an answer is reasonable.
Connections: U.S. System of Measurement; Geometry (substituting different sizes of baking pans while still maintaining the same area and/or volume)
Spin-offs: Have a party! Enjoy the rewards of accurate mathematical calculations.

Listed below are the ingredients in the recipe for a food to be used at a party. Use this recipe to help plan the party by answering the questions on this page. SHOW YOUR WORK. Simplify each fraction and use mixed numerals for fractions greater than 1.

SHRIMP APPETIZER

3	small (6-oz) cans shrimp	2	dashes hot sauce
1	egg, boiled and mashed	$3\frac{1}{2}$	tablespoons chopped fresh chives
$\frac{1}{2}$	cup mayonnaise	$\frac{1}{4}$	teaspoon salt
$2\frac{1}{2}$	tablespoons chopped onion	$\frac{1}{2}$	teaspoon chili powder
$3\frac{2}{3}$	tablespoons plain yogurt	1	tablespoon lemon juice

Rinse shrimp in cold water. Mash shrimp and cooked egg together. Add all remaining ingredients except chives. Stir well. Fold in chives. Chill and serve with chips or crackers. SERVES 6 PEOPLE.

1. What are the ingredients for $\frac{1}{2}$ of this recipe? Fill in the blanks in the recipe below.

_____	small (6-oz) cans shrimp	_____	dashes hot sauce
_____	egg, boiled and mashed	_____	tablespoons chopped fresh chives
_____	cup mayonnaise	_____	teaspoon salt
_____	tablespoons chopped onion	_____	teaspoon chili powder
_____	tablespoons plain yogurt	_____	tablespoon lemon juice

2. What are the ingredients for the recipe if 18 people attend the party? Make your own list of ingredients in the space below.

3. If each person drinks $2\frac{2}{3}$ cups of punch, how many cups of punch will be needed for 18 people?

Listed below are the ingredients in the recipe for another food to be used at a party. Use this recipe to help plan the party by answering the questions on this page.

APPLE CRUMBLE

4	cups apples, pared and sliced	$\frac{3}{4}$	teaspoon cinnamon
$\frac{3}{4}$	cup brown sugar	$\frac{1}{8}$	teaspoon allspice
$\frac{1}{2}$	cup dry oats	$\frac{1}{4}$	teaspoon nutmeg
$\frac{2}{3}$	cup flour	$\frac{1}{3}$	cup melted butter

Preheat oven to 375°. Place apples in greased 8-inch square pan. Blend remaining ingredients until crumbly and spread over the apples. Bake 30–35 minutes uncovered, until top is golden and apples are tender. SERVES 4 PEOPLE.

4. What are the ingredients for the recipe if 18 people attend the party? We do *not* want any leftovers! Fill in the blanks in the recipe below.

_____	cups apples, pared and sliced	_____	teaspoons cinnamon
_____	cups brown sugar	_____	teaspoons allspice
_____	cups dry oats	_____	teaspoons nutmeg
_____	cups flour	_____	cups melted butter

5. How many times would you need to fill a $\frac{2}{3}$ -cup container to measure 4 cups of apples?

6. If it takes $\frac{3}{4}$ cup of brown sugar to make one batch of apple crumble, how many batches could you make with 6 cups of brown sugar?

7. Exchange papers with your partner. Partner's name _____. Check your partner's work and compare answers. Be sure you both followed all of the directions for this assignment.

8. As you worked this assignment, how did you decide whether your answers were reasonable? (Use complete sentences in your answer.)

Activity 1-3: Stock Market

NOTES TO THE INSTRUCTOR

Summary: Information explaining stock market quotes is provided. Students determine ending prices for stocks using given data. Then each person chooses two real stocks and tracks their progress on the stock market. Possible profit or loss from the sale of these stocks is determined for each student and for the group as an investment team.

Skills Required: Fractions, Mixed Numerals
Skills Used: Addition and Subtraction of Fractions and Mixed Numerals, Stock Market Quotes, Group Investments

Grouping: Count Off (3 to 4 students per group)
Materials Needed and Preparation: Copies of Activity 1-3, one for each student; daily copy of a newspaper with stock quotes (or refer students to the library); copies of Activity A: *Think Teams* or Activity B: *What Do Employers Want?,* if needed, one for each student
Student Time: [**In Class**] 15 to 20 minutes on Day 1; 5 working days to track stock prices; 15 minutes to answer questions 4 and 5 on Day 2; [**Out of Class**] none, unless the library is used for stock quotes; OPTIONAL: individual homework assignment

Teaching Tips: Use Activity A or B to help students understand the value of working in groups. Discuss stock market quotes and refer students to quotes on television and in newspapers. Allow students to work in groups in class on the day the activity is introduced and on the day it is turned in. Tracking individual stocks can be done outside of class either in the library or by using a posted newspaper listing before or after class. The activity can be started any day of the week as long as students are allowed five working days to track the stocks. Question 5 may be used as extra credit or as part of the assignment. Emphasize the need to show work on both pages, then ask students to compare the process each of them used to answer question 1. For example, some students see patterns and use grouping techniques that can be used to illustrate the associative law of addition. OPTIONAL: Assign questions 1 through 3 as homework or as a take-home quiz, then do questions 4 and 5 in class. This reduces the amount of time spent in class on this activity. Also, question 5 may be used as extra credit.

Grading Tips: Individual Grade; 30 points for question 1; 10 points each for questions 2 and 3; 20 points for question 4; 30 points for question 5 (Adjust points if question 5 is used as extra credit.)

Comments: With the wonderful diversity found in developmental classes, it is possible to find a student who has active stock investments sitting next to a student who has never heard of the stock market. This activity encourages the students to share their knowledge and to begin building collaborative skills by working together on a direct application of math in the business world.
Connections: Positive and Negative Numbers, Associative Law of Addition

Activity 1-3: Stock Market   Name:_____

Basic Math   Fractions

Market prices for stocks are often quoted using fractions. The price of a stock can increase (go "up") or decrease (go "down"). For example, if a stock started at $31\frac{1}{2}$ per share and then went "up $\$\frac{3}{4}$," or $+\frac{3}{4}$, the new market price would be $\$32\frac{1}{4}$:

$$31\frac{1}{2} + \frac{3}{4} = 32\frac{1}{4}$$

If that same stock then went "down $\$\frac{1}{8}$," or $-\frac{1}{8}$, the market price would change to $\$32\frac{1}{8}$:

$$32\frac{1}{4} - \frac{1}{8} = 32\frac{1}{8}$$

1. The chart below shows the changes that occurred during one week. Find the ending price for each stock. Use the bottom of this page to show your work.

	Stock Alpha	Stock Beta	Stock Gamma
Starting Price:	$\$7\frac{1}{2}$	$\$26\frac{3}{8}$	$\$15$
Monday	up $1\frac{1}{4}$	no change	down $1\frac{5}{16}$
Tuesday	up $\frac{1}{8}$	down $\frac{7}{8}$	no change
Wednesday	down $\frac{3}{8}$	down $\frac{3}{4}$	up $1\frac{7}{8}$
Thursday	up $2\frac{1}{2}$	up $\frac{1}{2}$	up 4
Friday	down $\frac{1}{8}$	no change	no change
Ending Price:	_____	_____	_____

SHOW YOUR WORK HERE.

Stock market prices are shown daily in most newspapers and in the *Wall Street Journal.* Check the library for copies, or share copies in the classroom.

2. Select two stocks and track the market price of each stock for five working days. *Each person in your group should select different stocks.* Use the chart below to record the daily price and any changes up or down.

Stock Name _____   Stock Name _____

Day 1: Price: _____   Price: _____

Day 2: Price: _____ Change: _____   Price: _____ Change: _____

Day 3: Price: _____ Change: _____   Price: _____ Change: _____

Day 4: Price: _____ Change: _____   Price: _____ Change: _____

Day 5: Price: _____ Change: _____   Price: _____ Change: _____

3. If you had purchased 100 shares of these stocks on Day 1 and sold those shares on Day 5, would you have made money (a profit) or lost money (a loss)? How much money did you earn or lose on each stock? Show your work.

   a.   b.

4. a. Write the names of the other people in your group:_____

   _____

   b. Did the other members of your group show a gain or loss of money on their stocks? Show the results from question 3 for each person.

EXPLORE:

5. If your group had been pooling money to buy these stocks as a team, would the team have shown a profit or a loss after selling all of the stocks on Day 5? Use the results from questions 3 and 4 to find your answer. SHOW YOUR WORK!

NOTES TO THE INSTRUCTOR

Summary: Students use their own words to write step-by-step instructions for operations with fractions. The steps are group tested for usefulness and accuracy using sample problems.

Skills Required: Operations with Fractions
Skills Used: Writing Mathematical Procedures, Review of Operations with Fractions

Grouping: Count Off or Structured Groups, Ability Mixing (4 students per group)
Materials Needed and Preparation: Copies of Activity 1-4, one for each student
Student Time: [**In Class**] 30 minutes; OPTIONAL: 10 minutes on Day 1; 10 to 15 minutes on Day 2; [**Out of Class**] none; OPTIONAL: individual homework assignment

Teaching Tips: Allow students to read the directions printed on the activity before breaking into groups. Discuss the importance of listing all the steps that may be required to solve a problem or complete a task. For example, the directions printed on the activity show a step-by-step list of procedures to be followed in completing this activity. Encourage the students to use their own words for the steps used in operations with fractions, without referring to the textbook or class notes. Remind them to use words useful for solving *any* problem. For example, in problem 1, "2 times 3 plus 1" is very specific, but "multiply the whole number times the denominator and add the numerator" would be one useful step in the process of changing *any* mixed numeral to fractional notation. Emphasize the value of the final list of steps as a quick reference or "study sheet" for working with fractions. As the students work in groups, have them exchange papers and try to use the steps generated by their team members. Do not try to structure or limit the number of steps used for each operation. Allow students to list as many steps as they feel they need to achieve understanding and an accurate answer to the problem. The second problem in each operation may require a step not needed in the previous problem. This should create discussion and the rewriting of steps. OPTIONAL: Introduce the activity on Day 1 as a homework assignment to be completed in groups in class on Day 2.

Grading Tips: Individual Grade; 25 points for each operation; OPTIONAL: Do not grade but allow students to use the completed activity as a reference during a fractions quiz or test.

Comments: The value of this activity is its emphasis on encouraging students to use their own words to explain the steps involved in working with fractions. Although the use of proper mathematical vocabulary is a possible goal, it is not necessary for the instructor to stress this. It is amazing how hard students will work to perfect clarity of wording and usefulness of these steps if they know the activity can be used as a reference during a quiz or test!
Spin-offs: Allow students to use this activity as a reference during a quiz.

Directions:

1. Assign one person in your group to work on multiplication, one person to work on division, one on addition, and one on subtraction.

2. Solve the first problem in your assigned operation and check your answer.

3. Think about the steps you followed to obtain your answer. Write those steps in the space on the right. REMEMBER to use your own words and to use words that will help you solve *any* similar fraction problem. (Think of this as if you were preparing a "study sheet" on fractions!)

4. Exchange papers with your teammates. The other people in your group should be able to use your list of steps to solve the second problem in each operation.

5. Did your list of steps cover everything you needed to do? If not, work with your group to revise the steps.

Multiplication

1. $\dfrac{3}{8} \cdot 2\dfrac{1}{3} = \dfrac{7}{8}$

2. $6 \cdot 10\dfrac{2}{9} =$

Division

3. $12 \div 1\dfrac{4}{5} = 6\dfrac{2}{3}$

4. $1\dfrac{5}{12} \div \dfrac{3}{16} =$

Multiplication Steps:

Division Steps:

Follow the directions on page 1 of this activity.

Addition

5. $2\frac{1}{7} + 1\frac{2}{3} = 3\frac{17}{21}$

6. $4\frac{3}{5} + 2\frac{3}{20} =$

Subtraction

7. $9\frac{3}{8} - 2\frac{1}{3} = 7\frac{1}{24}$

8. $5\frac{1}{6} - 2\frac{1}{4} =$

Addition Steps:

Subtraction Steps:

NOTES TO THE INSTRUCTOR

Summary: This is a story of one day in the life of a student who needs to know whether he or she has enough money to be part of an exciting weekend opportunity. Bulletins throughout the day relate information about cash and bank account balances. Students use this information to decide whether they have enough money.

Skills Required: Addition and Subtraction
Skills Used: Addition and Subtraction of Decimals, Critical Thinking

Grouping: Any technique (3 to 4 students per group)
Materials Needed and Preparation: Copies of Activity 1-5, one for each group; prepare copies of hourly bulletins for each group by cutting the page as indicated.
Student Time: [In Class] 20 to 25 minutes; [Out of Class] none

Teaching Tips: Read the opening "9:15 A.M." statement in the story to set the stage for this activity. Form groups and ask students to designate one person from each group as a runner or messenger to receive information bulletins for the group. As each group completes calculations and answers questions, the runner will bring the activity to the instructor to receive the next bulletin. If the calculations are incorrect, the runner can be sent back so the group can try again. To encourage group accountability, announce that points will be deducted for each incorrect answer (or points added for correct answers). Two or more bulletins may be combined and given out together if you prefer to minimize physical movement in the classroom. For example, "11:00 A.M." and "noon" could be given out together, as well as "1:30 P.M." and "2:30 P.M." The groups may complete the calculations at different speeds, but the activity does not have to be a race. If you wish to eliminate competition between the groups, announce the bulletins to the entire class at regular intervals rather than using runners. There is no right or wrong decision at the end of the activity, although students should be encouraged to explain their decisions after looking at all of the issues involved. For example, the groups may want to consider the long-term effects of paying bank charges.

Grading Tips: Group Grade based on the accuracy of calculations (70 points) and the quality of the response to the final decision in the activity (30 points); OPTIONAL: Participation Grade for each group member

Comments: Some knowledge of checking accounts is useful but not required for this activity.

Activity 1-5: Do You Have Enough Money?   Name:_____

Basic Math _____ Decimals

This is a story of one day in your life.

9:15 A.M.   You have just learned of a great opportunity for this weekend and you really want to be part of it. You will have to rearrange things in your busy schedule, but that is okay. The big question is, do you have enough money? The tickets cost $32.50 and you will need $11.48 to cover transportation costs. You want about $15 for food and maybe $20 in case you decide to buy any souvenirs. You do *not* want to use a credit card.

Your friends need to know your decision by 5:00 P.M. today so they can buy the tickets. As the day progresses, you will get information that you can use with your group to figure out exactly how much money you have available. Use the space below to keep track of your money.

1. How much money would you need for this weekend?

Cash on Hand	**Bank Account Balance**

11:00 A.M.

noon

1:30 P.M.

2:30 P.M.

3:00 P.M.

4:00 P.M.   1.

2.

5:00 P.M.   Use the back of this page to
1. explain your decision and
2. list the names of the people in your group.

Instructor: The information on this page is given to the students in a special way. *Please read the Notes to the Instructor for Activity 1-5* for suggestions on using this information.

— — — — — — — — — — — — — —CUT— — — — — — — — — — — — — — —

11:00 A.M.   A quick search turns up $7 in cash plus 83¢ in change. Then, you look at your checkbook and find that you started this month with $35.21 in your account. You wrote checks for $22.50 and $7.38 and you deposited $42.00 into your account.
1. How much cash do you have on hand?
2. How much money is in your bank account?

— — — — — — — — — — — — — —CUT— — — — — — — — — — — — — — —

noon   You pay cash for your lunch, which costs $3.19, and one of your friends gives you $10.00 cash as a loan.

— — — — — — — — — — — — — —CUT— — — — — — — — — — — — — — —

1:30 P.M.   In your coat pocket you find two ATM (automated teller machine) receipts you had forgotten about. These receipts show that you took $16.00 and $25.00 out of your bank account.

— — — — — — — — — — — — — —CUT— — — — — — — — — — — — — — —

2:30 P.M.   You call your bank and learn that your paycheck for $234.76 has been deposited to your account, and the bank has taken $15 from your account for service fees.

— — — — — — — — — — — — — —CUT— — — — — — — — — — — — — — —

3:00 P.M.   Your mail comes with bills that you must pay immediately. You write checks for $17.99 and $43.58 to pay the bills. There is also a letter from your cousin that includes $20 in cash to repay the money you loaned him last month.

— — — — — — — — — — — — — —CUT— — — — — — — — — — — — — — —

4:00 P.M.   Your friends call to find out if you can go this weekend. You explain that if the balance in your bank account drops below $100, you will have to pay $15 to the bank as a service charge. Also, you have been thinking about next week. After the weekend is over, you will need to have about $60 left to live on until your next paycheck. You ask them to call you back in an hour.
1. If you go, will you have enough money until your next paycheck?
2. If you go, will you have to pay $15 to the bank as a service charge?

— — — — — — — — — — — — — —CUT— — — — — — — — — — — — — — —

5:00 P.M.   Your friends call again to find out if they should buy tickets for you for this weekend.
1. Explain your decision.
2. List the names of your group members.

NOTES TO THE INSTRUCTOR

Summary: Partners take turns being either the salesperson or the customer as they analyze four sales pitches.

Skills Required: Decimal Operations
Skills Used: Applications with Decimals, Critical Thinking

Grouping: Proximity Pairing (2 students per group)
Materials Needed and Preparation: Copies of Activity 1-6, one for each student; prepare copies of the sales pitches and cut the page as indicated. Use envelopes to enclose one copy of the information on the page for each set of partners.
Student Time: [**In Class**] 15 to 20 minutes; [**Out of Class**] none; OPTIONAL: question 11 can be assigned as individual homework

Teaching Tips: Use this activity after covering the decimal operations. Distribute the activity to each student and an envelope of sales pitch information to each set of partners. Remind students to take turns as salesperson and customer as they work together to answer questions 1 through 10. Question 5 (and question 11) can be used to generate a class discussion. Although question 11 is optional, it is particularly useful as a way to generate critical thinking.

Grading Tips: Individual or Group Grade; 10 points each for questions 1 through 10; extra credit for question 11, if it is used

Comments: As consumers, students need to use decimal calculations effectively to make informed spending decisions. This activity examines four common sales pitches and encourages students to think about all of the variables involved in each decision. Question 11 provides an opportunity for critical thinking as well as writing across the curriculum.
Spin-offs: Students can compare the sales pitches in the activity to the costs and conditions found in their own local area.

"Have we got a deal for YOU!" But is it really a good deal? How much money will you spend? How much money will you save? This activity has four examples of sales pitches. Take turns with your partner being either the salesperson or the customer as you analyze each example. SHOW YOUR WORK on each question. Partner's name_____

Health Club

1. If the customer only uses the club for six months, compare the cost of Plan A and Plan B.

   Cost for Plan A:

   Cost for Plan B:

   Which is less expensive?

2. If the customer uses the club for a year, how much would be saved by taking Plan C instead of Plan B?

More Music

3. In one year, how much would the customer spend for 5 compact disks at the discount store?

4. In one year, how much money would the customer spend for compact disks at More Music by meeting the membership requirements?

5. Would the customer save money by joining More Music? What other factors might influence the customer's choice?

Fine Furniture Rental

6. What would be the total cost for the customer to "rent to own" the table and chairs from Fine Furniture Rental?

7. How much would the customer pay for the table and chairs at the furniture store?

8. If the customer buys from the furniture store, how much money will be required for a down payment?

Telephone Shopping

9. What is the cost for each item

   a. if three items are purchased?

   b. if four items are purchased?

10. What other information does the customer need to know before deciding whether or not to place the order?

EXPLORE:

11. Examine the calculations from questions 6, 7, and 8. Write a paragraph comparing the cost of rent to own with the cost of buying from the furniture store. Explain what you would decide if you were the customer.

Instructor: The information on this page is given to the students in a special way. *Please read the Notes to the Instructor for Activity 1-6* for suggestions on using this activity and the information on this page.

— — — — — — — — — — — — — — —CUT— — — — — — — — — — — — — — —

Health Club

Salesperson: Join our club and become a better, healthier you! With Plan C you pay $153 up front and then $13.90 each month for the next 11 months.

Customer: I want to get the best deal for my money. Your advertisement also mentions that with Plan A, I just pay $90 to use the club for three months. Plan B is $27.50 per month for six months.

— — — — — — — — — — — — — — —CUT— — — — — — — — — — — — — — —

More Music

Salesperson: If you buy from More Music you'll save big money. When you become a member, your first four compact disks (CD's) will cost only $0.40 each and that includes tax and shipping charges. Then, within the same year you MUST buy at least two CD's. Each one costs $15.00, plus $2.00 for shipping and tax.

Customer: I usually buy about five CD's a year. I shop at a local discount store and pay $7.98 plus $0.39 tax for each one.

— — — — — — — — — — — — — — —CUT— — — — — — — — — — — — — — —

Fine Furniture Rental

Salesperson: Rent to own! With nothing down and no credit check, you can have a table and four chairs for just $53.93 per month. That includes tax and insurance. And in just 18 months the furniture is yours, free and clear!

Customer: I can buy a table and four chairs from a furniture store for $449.95 plus $28.40 tax. There are no interest charges if I pay one-third of the total cost as a down payment, and then finish paying for the furniture within 90 days.

— — — — — — — — — — — — — — —CUT— — — — — — — — — — — — — — —

Telephone Shopping

Salesperson: Use our toll free number! Operators are standing by to take your order. It's three items for $59.97 or four for $76.92.

Customer: Let me think about this. I can not spend more than $80.

NOTES TO THE INSTRUCTOR

Summary: Parts I and III of this activity allow student teams to determine ratios and solve proportions based on information available from their classmates. Part II provides students with the opportunity to learn ratio and proportion concepts by team teaching.

Skills Required: Ratios

Skills Used: Rates, Unit Rates, Proportions, Solving Proportions, Proportion Application Problems, Team Oral Presentations; OPTIONAL: Creating Application Problems

Grouping: Structured Groups

Materials Needed and Preparation: Copies of Activity 1-7, one for each student; copies of Activity A, Activity C, and Activity E, if needed, one for each student; plan Structured Groups or Teams; using page 2 of Activity 1-7, decide which problems to assign to which teams

Student Time: [**In Class**] 20 minutes on Day 1, full class period on Day 2 for presentations; OPTIONAL: 5 to 10 minutes per team presentation, spread over several class periods; [**Out of Class**] group homework assignment

Teaching Tips: This activity is designed to replace 90% of the lecture time normally used to cover ratios and proportions. Use Activity A and Activity C, if needed. Introduce the concept of ratios, and explain this activity before breaking into groups or teams. Assign the problems from page 2 so that each team is responsible for at least one problem. Problems from the textbook may be used instead of or in addition to the problems on page 2. Allow time on Day 1 for the students to do Part I. It is possible for ratios in Part I to contain a zero in the denominator, for example, if there are no part-time students in the class. If this occurs, instruct students to reverse the order of the ratio so zero is in the numerator. The groups will need to exchange information to answer question 2 in Part I.

On Day 1, the students should also make plans for Part II. They may need to meet as teams outside of class. Emphasize the importance of following the steps listed in Part II. Announce the date when teams should be ready to give their presentations. The presentations can be done in one class period or spread over several classes. Explain to students the relationship between learning a concept and explaining that concept to other people. Information on the number of "students on this campus" is needed for Part III, which can be completed either in class or as a homework assignment.

When the teams give their presentations, be alert for any inaccurate or incomplete information. Encourage the class to ask questions after each presentation. If the EXPLORE section is used, those student-generated problems will provide further reinforcement of the skills learned. After this activity is concluded, use Activity E: *How Are Things Going?* Discuss the results and ask for suggestions for improving future group assignments.

NOTES TO THE INSTRUCTOR

Grading Tips: Group or Combined Grade; 5 points for the names of team members and 5 points for each answer in Parts I and III for a total of 50 points; the remaining 50 points to come from the team presentation in Part II; OPTIONAL: If the EXPLORE section is used, allow 30 points for the presentations and 20 points for the EXPLORE problems.

Comments: "If you really want to *learn* something, you should *teach it*." Instructors know the validity of this statement. This activity allows students to learn by teaching. This approach is appropriate here for several reasons: (1) The concepts in ratio and proportion are generally not too threatening to most students; (2) these topics are usually covered several weeks after the beginning of classes, so the students can feel comfortable presenting information before the class; (3) the wealth of examples of ratios, rates, and proportions in daily life provides students with opportunities to create their own application problems; (4) yes, developmental math students *can* teach these concepts!

Connections: This "learning through teaching" approach can be used successfully with any or all of the concepts in developmental basic math.

Spin-offs: Prepare a homework assignment or class quiz, using the student-generated problems from the EXPLORE section of this activity.

Activity 1-7:   Making Comparisons   Name:_____

Basic Math   Ratio and Proportion

Work together with your team to complete this activity. You will need to use time in class and out of class to finish each part.

Team Members: _____

_____

Part I: Show your work on each problem, and SHOW ALL RATIOS IN SIMPLEST FORM.

1. Find these ratios for the students in your team:

   a. the ratio of full-time students to part-time students _____

   b. the ratio of women students to men students _____

   c. the ratio of left-handed students to right-handed students _____

2. Find these ratios for the students in this math class:

   a. the ratio of women students to total students in this class _____

   b. the ratio of full-time students to part-time students _____

   c. the ratio of left-handed students to total students _____

Part II: Your instructor will assign certain ratio and proportion problems to your team. Work as a team to learn how to solve these problems and how to explain them to the rest of the class.

1. With your team members, match your assigned problems to the section of the textbook that covers similar problems.

2. Read this section carefully and talk about it with your team. Work some of the exercises in the textbook and check your answers together.

3. Decide with your team how you will work together to explain your assigned problems to the rest of the class. What steps are needed to work these problems?

4. Be ready with your team to talk about your assigned ratio and proportion problems with the class.

Part III: Assume that the ratios you found in Part I for this math class are true for all the students on this campus. Use this assumption to answer the following questions:

1. How many students on this campus are left-handed?

2. How many students on this campus are women?

3. How many students on this campus are full-time students?

Your instructor will assign your team one or more of these problems or similar problems from the textbook.

1. A student spent $200 for books and $2400 for tuition. (Show all ratios in simplest form.)

   a. What is the ratio of the amount spent for books to the amount spent for tuition?

   b. What is the ratio of the amount spent for books to the total amount spent by the student?

2. Determine which of the following number pairs or ratios are proportional and which are not.

   a. 5, 8 and 15, 24
   b. 3, 4 and 54, 72

   c. $\dfrac{3}{18}$ and $\dfrac{4}{19}$
   d. $\dfrac{16}{3}$ and $\dfrac{48}{9}$

3. Solve. Round to the nearest tenth, if necessary.

   a. $\dfrac{8}{6} = \dfrac{x}{15}$
   b. $\dfrac{0.3}{5.4} = \dfrac{n}{41}$

4. Solve.

   a. $\dfrac{21}{n} = \dfrac{6}{14}$
   b. $\dfrac{3}{8} = \dfrac{3\frac{3}{4}}{y}$

5. A jogger ran 7.5 miles in 50 minutes. What was the rate in miles per minute?

6. A 15-oz can of beans costs $0.59. A 28-oz can of beans costs $1.19. Which has the lower unit price?

7. To make liquid food for a hummingbird, you need a ratio of 1 cup of sugar to 4 cups of water. If you use $1\frac{1}{2}$ cups of water, how much sugar will you need?

8. A plant fertilizer is made by mixing 2 tablespoons of powder with 3 gallons of water. How many gallons of water should be mixed with 5 tablespoons of powder?

EXPLORE:

9. Work with your team members to create 2 new problems similar to the problems assigned to your team. Use your textbook, newspaper articles, or your own experiences to find ideas for these problems. The problems your team creates may be used as part of a homework assignment or quiz to be completed by the class. Be prepared to answer questions and to help your classmates work your new problems.

   a. Write the new problems your team has created.

   b. Show how to solve each problem and the correct answer for each problem.

NOTES TO THE INSTRUCTOR

Summary: The relationships between percents are visually represented using lines. Students use the lines to answer questions and then create lines and/or diagrams of their own.

Skills Required: None, Fractions helpful
Skills Used: Percents, Comparing Percents

Grouping: Structured Groups, Ability Mixing, or Count Off (3 to 4 students per group)
Materials Needed and Preparation: Copies of Activity 1-8, one for each student; copies of Activity C: *To the Student,* one for each student, if needed; plan Structured Groups, if necessary.
Student Time: [**In Class**] 5 minutes on Day 1, 10 to 15 minutes on Day 2; [**Out of Class**] individual homework assignment

Teaching Tips: Use this activity to introduce percents. On Day 1, explain the meaning of percents and draw a line on the overhead or board to use as an example of the lines used in this activity. Assign page 1 of the activity as homework. If the EXPLORE section is used, ask students to look for examples of percents in newspapers and magazines. On Day 2, distribute Activity C, if needed, and have the students work on page 2 in groups. The answers to question 12(b and d), on the second page, can be found *before* the concept of translating percent statements into algebra has been introduced, by using visual estimation and/or knowledge of fractions. When the activity is completed, each group should turn in a group report containing the answers to page 2 along with each individual student's answers to page 1. OPTIONAL: The EXPLORE section may be deleted or used as extra credit.

Grading Tips: Combined Grade; 3 points for each of the 20 answers (including names of group members) to questions 1 through 16; 10 points for each part of the EXPLORE section; OPTIONAL: Readjust points if the EXPLORE section is deleted or counted as extra credit.

Comments: Developmental students often have difficulty seeing relationships between percents. They may not perceive any difference between 0.5% and 50%. And they may lack a reference for amounts over 100%, such as 350%. This activity gives the students an opportunity to see a visual representation of percents using simple lines.
Connections: Lines are generally used to illustrate concepts in geometry and in positive and negative numbers.
Spin-offs: Save the magazine and newspaper examples of percents from question 17. After the students have learned more about percents, they can create application problems based on these examples.

Basic Math   Percents

Note: Each line is marked off and labeled in equal parts.

First line: _____
             A   B   C   D   E

1. Start at letter A in the line above. If you go 80% of the way to letter E, which two letters are you between?

2. Start at letter A in the line above. If you go 29% of the way to letter E, where are you on the line? What percent is the rest of the line?

Second line: _____
              R   S   T   U   V   W

3. Start at letter R and go 250% *of the distance* to letter S. Which two letters are you between?

4. Start at letter R and go 400% *of the distance* to letter S. Where are you on the line?

Third line: _____
             G   H   I   J   K   L

Start at letter G. Mark the location of each person on the third line.

5. Maria has 52% of the line.

6. Adam has 4% of the line.

7. Chan has 73.9% of the line.

8. How would you show that Sharla has 150% of the line?

Look at the four lines on the right.
Line A shows 10%.

9. What percent does line B show?

10. What percent does line C show?

11. What percent does line D show?

A ———————+——————— 10%

B ——————+—— ?

C —+— ?

D — ?

It's Your Turn

Decide with your group how you will answer the questions on this page. You will turn in *your individual answers* to page 1, along with *your group's report* answering the questions on page 2.

Group Members: _____

_____

12. The bottom of this page measures $8\frac{1}{2}$ inches.

   a. Use a line or marks to show 50% of the bottom of this page.

   b. What is 50% of $8\frac{1}{2}$?

   c. How could you show 200% of $8\frac{1}{2}$?

   d. What is 200% of $8\frac{1}{2}$?

13. Sales tax is 6%. Show 6% of $1.00 using a line or a diagram.

14. Use a line or diagram to show 1%.

15. Use a line or diagram to compare 50%, 5%, and 0.5%.

16. Look at question 11 on page 1 of this activity. Turn it upside down and look at the relationship between the lines. *If line D represents 100%,* what percent is line B?

EXPLORE:

17. Use magazines and newspapers to find examples of the following uses of percents in our daily lives. Copy or cut out each example and turn it in with your group report.

   a. Using percents for a sale or discount.

   b. Using percents to report results or show a trend.

   c. Using percents when money is loaned or borrowed.

   d. Using percents to show an increase or decrease.

NOTES TO THE INSTRUCTOR

Summary: Students role play and use percent concepts as they interview each other to learn enough information about the effects of a storm on a city to write a news report.

Skills Required: Basic Percent Equation, Percent of Increase and Decrease
Skills Used: Sales Tax, Simple Interest, Commission, Discount, Percent of Increase, Percent of Decrease

Grouping: Any technique (5 students per group)
Materials Needed and Preparation: Copies of page 1 of Activity 1-9, one for each student. Also, prepare copies of page 2 of the activity, which has the roles students will play. You will need one copy of page 2 per group, cut along the lines so that each student has at least two roles.
Student Time: [**In Class**] 20 to 30 minutes; [**Out of Class**] individual homework assignment if question 11 is used

Teaching Tips: This activity may be used as percent concepts are taught, or as a review to reinforce learning before an exam. Before forming groups, explain that each student will have at least two roles to play. The roles contain key information that is needed to answer the questions in the activity, so the people in each group will need to interview each other and to share the information. Distribute the roles from page 2 so each group has a full set of the ten roles in the activity. OPTIONAL: Question 11 expands the activity to include writing a news report. This section may be omitted, or given extra credit.

Grading Tips: Individual or Combined Grade; 8 points each for questions 1 through 10, and 20 points for question 11; readjust points if question 11 is omitted.

Comments: This is a real-world activity based on the fact that storms and natural disasters make the news every day. Those news reports contain numbers, facts, and percentages that help the audience understand the event. Therefore this activity combines mathematics with people's natural curiosity to learn more about what is in the news.

Basic Math   Percents

A severe storm has hit the city of Toobad in your state. There has been property damage and many people have been forced to leave their homes. You are a reporter sent to the city to report on this storm and its effect on the people. Your information must be accurate because your report will be used as part of the national news.

To get the facts about this storm, you will interview several people in Toobad. Each person in your group has one or two roles to play and the information you will need to write your report. First, use the information from your group members to answer the following questions:

1. Find the population of Toobad by talking to the MAYOR.

2. Use information from the FARMER to determine what percent of the acres planted with crops are underwater.

3. Ask the GOVERNOR how much money the state will send to help clean up the damage from the storm.

4. Talk to the U.S. SENATOR and the BUSINESS OWNER. If the business owner borrows the money needed to rebuild from the federal government at a low interest rate, how much will the interest be for 1 year?

5. What is the percent of decrease in the value of the HOMEOWNER's house?

6. If the INSURANCE AGENT had sold a policy worth $80,000 last week, how much commission would he or she have received?

7. What is the total cost each night for the TOURIST to stay in the motel if the tax is 9%?

8. Talk to the EMERGENCY SHELTER WORKER. How many people are expected to be in the emergency shelters tonight?

9. If the STORE OWNER has items that usually cost $45 each, how much will each item cost during the "Storm Sale?"

10. If the U.S. SENATOR can increase government aid by 15%, how much more money would be made available to help rebuild Toobad?

11. Now, write a news report about the effects of this storm on the city of Toobad. Use the facts from your interviews and from your calculations in questions 1–10. State whether the report is to be used for television, newspapers, or a news magazine.

Instructor: The information on this page is given to the students in a special way. *Please read the Notes to the Instructor for Activity 1-9* for suggestions on using this activity and the roles on this page.

— — — — — — — — — — — — —CUT— — — — — — — — — — — — — — — —

MAYOR: "Right now, 39,400 people who normally live in Toobad have been forced to leave the city. That is 40% of the population of Toobad. Our city has been hit hard by this storm, but we will bounce back!"

U.S. SENATOR: "The federal government will release $2.4 million dollars in aid to help rebuild Toobad. Washington will offer low interest loans at 6% simple interest to help rebuild damaged businesses."

— — — — — — — — — — — — —CUT— — — — — — — — — — — — — — — —

BUSINESS OWNER: "My business was badly damaged. It would take at least $200,000 to rebuild it."

HOMEOWNER: "My house was worth $105,000 before this storm hit. Now I would be lucky to sell it for $63,000."

— — — — — — — — — — — — —CUT— — — — — — — — — — — — — — — —

FARMER: "I had 300 acres planted with crops. Now 75 of those acres are underwater."

TOURIST: "We sure didn't plan to spend our vacation like this! We are about to run out of money. We are stuck in a motel that costs $95.00 a night, plus tax."

— — — — — — — — — — — — —CUT— — — — — — — — — — — — — — — —

STATE GOVERNOR: "Our state disaster aid fund contains $500,000. We estimate that we will need to use 35% of that fund to clean up the damage from this storm."

STORE OWNER: "What a mess! Many items in my store are damaged. I'm going to have a 'Storm Sale' and discount everything 60%."

— — — — — — — — — — — — —CUT— — — — — — — — — — — — — — — —

EMERGENCY SHELTER WORKER: "We had 2,530 people in our emergency shelters yesterday. We expect an increase of 70% for tonight. I hope we have enough beds and blankets."

INSURANCE AGENT: "Last week I was selling insurance policies and earning 8% commission on every sale. This week I'm helping my policy holders file claims for damage from the storm."

NOTES TO THE INSTRUCTOR

Summary: Cumulative review problems are arranged so that the answer from each problem in a set becomes part of the next problem. Students review together by working problems and checking answers as the problem sets rotate around the group.

Skills Required: Whole Numbers, Order of Operations, Fractions, Decimals, Ratio and Proportion, Percents
Skills Used: Cumulative Review of skills listed above

Grouping: Any technique (4 students per group)
Materials Needed and Preparation: Copies of Activity 1-10, one for each student
Student Time: [**In Class**] 20 to 30 minutes; OPTIONAL: 5 minutes on Day 1, 10 to 15 minutes on Day 2; [**Out of Class**] none; OPTIONAL: individual homework assignment

Teaching Tips: This activity can be used to review computational skills prior to a midterm exam. Before the groups are formed, read aloud the directions on page 1 of the activity. Emphasize the need to check the work carefully on each problem, because the answer from one problem becomes part of the next problem. Distribute the activity sheets so the students can see how various symbols are used to link the answer from one problem to a specific part of the next problem in each set. *Ask students to show any fractional answers in simplified form.* Since the Final Challenge! questions require computations using the last answers from each set, it may be helpful to show the answers to question 4, for each set, on the board or overhead. OPTIONAL: On Day 1, explain the activity and assign it as individual homework. On Day 2, group students so they can compare answers and discuss questions.

Grading Tips: Participation Grade; OPTIONAL: Group Grade with 5 points for each problem in each set; 10 points for each problem in the Final Challenge!

Comments: Because this activity provides a cumulative review, it encourages students to synthesize skills and concepts as they prepare for a midterm exam.
Connections: Activity F: *Using College to Reach My Goals* is also useful as a midterm exercise.

Activity 1-10: Midterm Madness

Name:_____

Basic Math   Midterm Review

Directions: Each person in the group will start working on a different set of problems. Notice that the answer from each problem should be written in as part of the next problem in the set. Use the lines and symbols as a guide. When you finish one problem, pass your paper to the left so your teammate can do the next problem in the set. When a new problem is passed to you, be sure to check the work of the person ahead of you. As the problems rotate around the group, discuss any errors or questions with your teammates.

Review Set 1

1. $1421 \div 20.3 = $ _____

2. 30% of _____ $= \bigcirc$

3. $\dfrac{4}{\bigcirc} \div 6\dfrac{4}{9} = \square$

4. Solve for x. Round to the nearest tenth. $\dfrac{5}{x} = \square$

Review Set 2

1. $6\dfrac{1}{4} \cdot 1\dfrac{3}{5} = $ _____

2. $\dfrac{}{21} + 1\dfrac{6}{7} + 4\dfrac{2}{3} = \triangle$

3. Solve for n. $\dfrac{n}{5} = \dfrac{\triangle}{8}$

   $n = \langle\!\rangle$

4. $73 + 6.81 + 0.042 + \langle\!\rangle = $

Review Set 3

1. Solve for x. $\dfrac{x}{11} = \dfrac{32}{4}$

   $x = \triangle$

2. $5^2 + 6 - \triangle \div 8 = \bigcirc$

3. 40% of what is $\bigcirc$? _____

4. The LCM of _____, 6, and 30 is

Review Set 4

1. $98,041 - 97,969 = \langle\!\!\!\!\!\bigcirc\!\!\!\!\!\rangle$

2. $\langle\!\!\!\!\!\bigcirc\!\!\!\!\!\rangle$ is what percent of 200? _____

3. _____ written in fractional notation is $\square$

4. $\square \cdot \left(\dfrac{5}{9}\right)^3 =$

Final Challenge!

1. Which is larger, the last answer from Set 2, or 50% of the last answer from Set 1?

2. Multiply the last answer from Set 4 by the last answer from Set 3. Write the product as a mixed numeral.

NOTES TO THE INSTRUCTOR

Summary: Students work both individually and in groups to design and conduct surveys. Then they analyze the individual results and the combined group results as they prepare tables, draw graphs, and write a report.

Skills Required: Percents

Skills Used: Survey Methods, Data Analysis and Reporting, Tables, Circle Graphs, Pictographs, Bar Graphs, Report Writing

Grouping: Structured Groups, Ability Mixing (3 students per group)

Materials Needed and Preparation: Copies of Activity 1-11, one for each student; copies of Activity A: *Think Teams,* or Activity B: *What Do Employers Want?,* and Activity D: *Roles for Groups,* one for each student, if needed; plan Structured Groups.

Student Time: [In Class] 10 to 15 minutes each on Days 1 and 2; **[Out of Class]** individual and group homework assignment

Teaching Tips: This activity can be used to replace the lecture and laboratory time normally spent on pictographs, tables, circle graphs and bar graphs. Before Day 1, use Activity A or B, if you have not already done so, to explain the value of group work in preparing to meet employer expectations. Also, distribute Activity 1-11, and assign reading of the textbook material on statistics, as homework. On Day 1, use Activity D to help students understand the steps involved in working together on a long-term assignment. (Two of the roles in Activity D may be combined.) Form groups and ask each group to decide on a survey question and possible choices for the answer. You may want to approve the survey questions. Students complete the surveys outside of class and should be given several days or a week to complete the task. On Day 2, let the groups meet again to combine data and to assign graphs for question 8. Allow two or three additional days to complete the final survey reports. Remind students of the need for quality and neatness as they prepare the reports. They can use a computer to design and draw the graphs. Ask for verbal status reports as a way to check on the progress of each group.

Grading Tips: Individual, Group, or Combined Grade; 25 points for the written report (10 to 15 of these points should be for quality); 5 points for question 9(b); 10 points for question 9(c); 20 points each for question 9(d), 9(e), and 9(f).

Comments: During the time span of a long-term activity, such as "The Survey," attendance patterns in the classroom can be erratic. It is also possible for individual students to show varying degrees of responsibility as they work together to complete the group assignment. This activity compensates for these factors by providing both individual and group accountability. Students can be graded on their individual efforts, as well as their contributions to the group effort, without being penalized if other members in their group do not complete the assigned task. This activity also provides students with an opportunity to learn what is required to prepare a quality written report.

Spin-offs: Interesting survey results may be shared throughout the school by publishing reports in a student newsletter or by posting graphs on bulletin boards.

How many people prefer a certain brand of soft drink? What issues will be important to the voters in the next election? By questioning, tabulating, analyzing, and reporting, we can use statistics to describe the answers to these questions.

Read the information in your textbook about statistics and look at the graphs. Then work with your group members to complete this activity. *Be sure you know the date when this assignment is due to be turned in.*

Questioning

1. Decide on a question of interest to your group. The question should have *at least* five possible choices for the answer. If necessary, one of the choices can be "other." For example, "What is your favorite ice cream? (1) vanilla (2) rocky road (3) chocolate (4) strawberry (5) cookies and cream (6) other."

2. Using the question and list of choices selected by your group, each group member will conduct a survey of at least 20 people.

Tabulating

3. Make a table showing the responses of the people you interviewed.

4. Combine the responses from your survey with the responses your group members found in their surveys. Show the combined results from your group in another table.

Analyzing

5. Using your individual data, compute the percent of those interviewed who selected each choice.

6. Using the combined data from your group, compute the percent of those interviewed who selected each choice.

Reporting
Note: Each graph should have a title and show correct labels. Look at examples in your textbook.

7. Draw a circle graph that shows the results of your individual survey. Use a circular object, such as a cup, to help draw the circle.

8. Use graphs to show the combined results from your group. Each person in your group will draw a different type of graph. Decide which person will do which of the following graphs:

   a. pictograph

   b. bar graph

   c. circle graph

9. Prepare your individual report. At least 10 to 15 points of your grade will be determined by the *quality* of your presentation. Use complete sentences in your writing. Check your work to be sure it is neat and well organized.

   Begin your report with a brief written summary describing your survey. What question did you ask and what were your findings? How did your individual results compare with the combined results of the group?

   Include the following information in your report:

   a. brief written summary   (25 points, including quality of presentation)

   b. names of the other people in your group   (5 points)

   c. two tables of data from questions 3 and 4   (10 points)

   d. calculations of percentages from questions 5 and 6   (20 points)

   e. the circle graph you drew for question 7   (20 points)

   f. the graph you drew for the group in question 8   (20 points)

10. Turn in your report, along with the reports of the other members in your group, in one folder.

Activity 1-12: Using Measurement

NOTES TO THE INSTRUCTOR

Summary: Common classroom items, including rectangular and circular shapes, are measured using both U.S. and metric units. These measurements are then applied to questions involving perimeter and area.

Skills Required: Vocabulary for Measurement, Perimeter and Area
Skills Used: U.S. and Metric Measurement, Perimeter, Circumference, Diameter, Area

Grouping: Structured Groups, Ability Mixing
Materials Needed and Preparation: Copies of Activity 1-12, one for each student; U.S. and metric measuring devices for length, for each group; a few of the following items to be shared among the groups: U.S. quarters, college ruled notebook paper, same-size rubber bands, and scissors; plan Structured Groups.
Student Time: [**In Class**] 15 to 20 minutes for page 1; OPTIONAL: 20 to 30 minutes for page 2; [**Out of Class**] individual homework assignment; OPTIONAL: none

Teaching Tips: This activity is designed to replace 90% of the lecture time normally spent on linear measures. Introduce vocabulary for U.S. and metric measurement, and for perimeter and area. Form groups and distribute measuring devices and items to be measured. Suggest cutting the rubber band as a way to measure the circumference for question 6. After the measurements have been completed, ask each group to share its results. Guide the class in determining standard measurements for each item. These measurements will be used, as an individual homework assignment, for the calculations on page 2. Emphasize the detailed directions for "SHOW YOUR WORK" on page 2. Rounding directions may be necessary for question 15, depending on the size of the rubber band. OPTIONAL: The groups can complete the page 2 calculations during class time, so the entire activity could be completed during one class period. OPTIONAL: Page 2 can be used as a stand-alone activity without page 1. In this approach, the student groups decide what measurements are needed to work problems 8 through 16. Also, ask the students to define perimeter, area, and circumference in their own words and then to agree on a group definition for each term.

Grading Tips: Individual or Group Grade; 4 points each for questions 1 through 7; 8 points each for questions 9 through 16 (Partial credit may be given on page 2 by allowing 2 points for each of the four steps required in "SHOW YOUR WORK" for each problem.)

Comments: This is a hands-on activity that enables students to use, and to compare, both systems of linear measurement. The use of typical classroom items provides useful visual references for comparative sizes, such as inches and centimeters. After the students have completed the perimeter and area calculations, it is interesting to lead a class discussion on the relative merits of the two systems of measurement.
Connections: Volume measurements in Activity 1-13: *A Moving Experience*
Spin-offs: Extend this activity to include measurements of the walls and floor of the math classroom. A 25-foot metal tape measure can be used, or use a piece of string to determine the length and then measure the string.

DIRECTIONS: Use U.S. and metric units of length to answer the questions in this activity. Directions are given for each measurement. After the measurements are completed on page 1, each group will share their results with the class. The class will then determine the correct measurements to use as a standard for the calculations on page 2.

1. Measure the length and width of the front cover of the math textbook

    a. to the nearest eighth of an inch: length _____ width _____

    b. in centimeters: length _____ width _____

2. Measure the length and width of the sheet of college ruled notebook paper

    a. to the nearest eighth of an inch: length _____ width _____

    b. in centimeters: length _____ width _____

3. Measure the distance between the lines on the sheet of college ruled notebook paper

    a. to the nearest eighth of an inch: _____

    b. to the nearest millimeter: _____

4. Measure the distance from the top of the notebook paper to the first line

    a. to the nearest eighth of an inch: _____

    b. to the nearest millimeter: _____

5. Measure the diameter of a U.S. quarter coin

    a. in millimeters: _____

    b. in centimeters: _____

6. Measure the circumference of the rubber band

    a. to the nearest eighth of an inch: _____

    b. to the nearest millimeter: _____

7. Measure the length and width of the chalkboard or dry erase board in this classroom

    a. to the nearest inch: length _____ width _____

    b. to the nearest centimeter: length _____ width _____

DIRECTIONS: Use fractional notation for the U.S. units and decimal notation for the metric units. **SHOW YOUR WORK** on each problem. (1) Draw a picture and include the measurements. (2) Write the formula(s) you need to solve the problem. (3) Use the formula(s) to solve the problem. (4) State your answer with the appropriate units.

8. Find the perimeter, in inches, of the front cover of the math textbook.

9. Use centimeters to find the area of the front cover of the math textbook.

10. What is the perimeter, in centimeters, of the sheet of notebook paper?

11. Find the area, in U.S. units, of the sheet of notebook paper.

12. What is the area, in U.S. units, of the unlined space at the top of a sheet of notebook paper?

13. Find the circumference of a U.S. quarter, in millimeters. Use 3.14 for π.

14. Using centimeters, find the area of a U.S. quarter. Use 3.14 for π.

15. If the rubber band is formed into a circle, calculate the diameter of the circle, in millimeters. Use 3.14 for π.

16. Using U.S. units, find the perimeter of the chalkboard or dry erase board in this classroom.

NOTES TO THE INSTRUCTOR

Summary: Students use basic math and geometry concepts to make decisions relating to a household move. Information from U-Haul is used as a basis for packing and cost calculations.

Skills Required: Decimals, Reading a Chart, Calculators
Skills Used: Volume, Area, American Units of Measure, Critical Thinking

Grouping: Structured Groups
Materials Needed and Preparation: Copies of Activity 1-13 (do not copy page 2 on the back of page 1), one for each group; copies of Activity A: *Think Teams,* and Activity C: *To the Student,* if needed; calculators, one for each student or several for each group. Prepare copies of the decisions for each group by cutting page 3 as indicated. Plan Structured Groups.
Student Time: [**In Class**] 30 to 40 minutes; OPTIONAL: 15 to 20 minutes for the first and second decisions; [**Out of Class**] none; OPTIONAL: group homework assignment for the third, fourth, and fifth decisions

Teaching Tips: Use Activities A and C, if necessary. Read the opening statement in the story to set the stage for Activity 1-13. Form groups and ask students to designate one person from each group as a messenger to receive the information needed for each decision. As each group completes calculations and makes a decision, the messenger will bring the activity to the instructor to receive the information for the next decision. If the calculations are incorrect, the messenger can be sent back so the group can try again. The groups may complete the calculations at different speeds, but the activity does not have to be a race. For question 3, you may choose to provide the conversion information (1728 cubic inches = 1 cubic foot) to the class. There will be slight variations in the answer for question 3, depending upon the approach used for the calculations. Question 4 has two responses based on the total cargo of boxes plus 290 cubic feet of other furnishings. Question 10 is open-ended and may be used as a basis for a class discussion. OPTIONAL: Class time can be shortened by assigning the third, fourth, and fifth decisions as group homework.

Grading Tips: Group Grade; 10 points for each of the 10 questions; OPTIONAL: Participation Grade

Comments: A do-it-yourself move is a common event, one most people have experienced at least once. To reduce the amount of time required to complete this activity, some of the variables that are found in an actual move have been eliminated or streamlined. However, this activity still closely parallels the moving experiences of many people. Question 10 allows students to draw on their own backgrounds as they share useful information about moving expenses.

It's time to get moving! You've found a great place to live, so you must pack up and move. The last time you moved, it was easy. But since then you have accumulated more furniture and kitchen items. You will need to pack dishes, clothes, books, and so forth into boxes, and then load everything into a truck or a trailer for the drive to your new residence. Since you want to do the move yourself, you contact U-HAUL, a leading truck and trailer rental company, and receive the information shown on page 2.

As you get ready to move, there will be several decisions you can make together with your group. Use the space below to record the names of your group members and to show the group work for each decision. Show the formulas and solution steps you use, as well as the answer for each question.

Group Members: _____

_____

First Decision:
1.   2.

Second Decision:
3.   4.

Third Decision:
5.   6.

Fourth Decision:
7.   8.

Fifth Decision:
9.   10.

Use the information below as a work sheet to help you make your decisions.

NUMBER OF BOXES NEEDED FOR EACH ROOM

U-HAUL	Living Room	Dining Room	Kitchen	Per Bath-room	Per Bedroom x Quantity	(Misc.) Garage/ Storage	TOTAL BOXES	TOTAL PRICE
☐ **Small Box** 16" x 12" x 12" $1.25 1.33 cu/ft	2	1	2	1	1	2		
☐ **Medium Box** 18" x 18" x 16" $1.95	1	1	2	2	2	2		
☐ **Large Box** 18" x 18" x 24" $2.25	1		1		1	1		
☐ **Extra-Large Box** 24" x 18" x 24" $2.95	1				1			
☐ **Dish Pak™ Box** 18" x 18" x 28" $3.95			1					
☐ **Dish Saver** 24" x 12" x 12" $9.95			1					

Courtesy of U-HAUL International, Inc. Reprinted with permission.

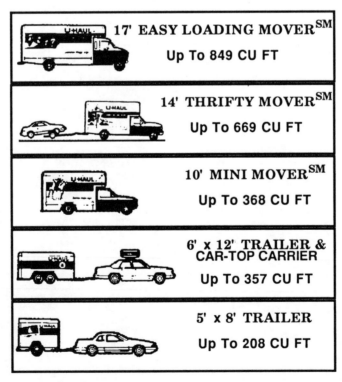

Courtesy of U-HAUL International, Inc. Reprinted with permission.

Instructor: The information on this page is given to the students in a special way. *Please read the Notes to the Instructor for Activity 1-13* for suggestions on using this activity and the information on this page.

————————————————————CUT————————————————————

First Decision: You need boxes. You must pack items from one bedroom, a living room, a kitchen, and a bathroom. There is no garage. Use the chart from U-HAUL as a work sheet to help you make these decisions.

   1. How many boxes of each size should you buy?

   2. What is the total cost for the boxes?

————————————————————CUT————————————————————

Second Decision: Now decide which truck or trailer to rent. The capacity for each one is given in cubic feet. You will be loading the packed boxes plus approximately 290 cubic feet of other furnishings.

   3. How many *cubic feet* of space will be needed for the packed boxes?
      Caution: The measurements for each box are given in inches. How many cubic inches = 1 cubic foot? (The conversion for a small box has been done for you.)

   4. Which truck could you rent? Which trailer could you rent?

————————————————————CUT————————————————————

Third Decision: Several of your friends volunteer to help you move. In return, they expect you to feed them! You can order two medium round pizzas, each 12 inches in diameter, for $11.34. Or you can order one rectangular pizza, which is 12 inches by 24 inches, for $11.68.

   5. Find the area of the two round pizzas and the area of the rectangular pizza. Use 3.14 for π.

   6. Which would be the better buy? (Which is the cheapest per square inch?) Round to the nearest cent.

————————————————————CUT————————————————————

Fourth Decision: One of your friends has a mini-van, and she offers to let you use it for the move. Because of time limitations, you can use the van for only one trip. The open space in the back of the van measures $5\frac{1}{2}$ feet long, 5 feet wide, and $4\frac{1}{2}$ feet high.

   7. How many cubic feet of space is available in the van?

   8. If you use her van, then which truck or trailer would you rent?

————————————————————CUT————————————————————

Fifth Decision: You call and ask U-HAUL about prices for the vehicle you have chosen to rent. The 10′ Mini Mover will cost $47.25. The 6′ × 12′ Trailer will cost $25.00.

   9. What will be your cost for boxes, plus pizza, plus the rental vehicle?

  10. What other costs should you plan for as part of this move?

NOTES TO THE INSTRUCTOR

Summary: Students apply geometric concepts to the design of a school park.

Skills Required: Geometry Formulas
Skills Used: Perimeter, Area, Volume, Pythagorean Theorem

Grouping: Any technique
Materials Needed and Preparation: Copies of Activity 1-14, one for each student
Student Time: [**In Class**] 15 to 20 minutes; [**Out of Class**] none; OPTIONAL: use the activity as a homework assignment

Teaching Tips: Use this activity to review and to consolidate concepts after covering the basic geometric formulas for perimeter, area, volume, and the Pythagorean Theorem. Introduce the activity and form groups. Announce whether you expect a group or an individual report for Part II. Students will need to apply several formulas to answer the questions in Part I. There are no right or wrong answers for the decisions on where to locate the various design elements, such as the fish pond, in the park. OPTIONAL: Use this activity as a homework assignment to help review for a test. The activity can be extended by asking the students to determine costs for fencing and gates and for grass sod or seed for the park. Students can also use a computer to design and draw the planned park.

Grading Tips: Group or Individual Grade; 15 points each for questions 1 through 6; 10 points each for Part II, a and c; OPTIONAL: Participation Grade

Comments: Because this activity provides a good review of the basic geometric formulas, it can be used to prepare for a test or an exam. The design and drawing portions of the activity may appeal to students who excel in areas other than math.

A piece of land, to be used as a park, has been donated to the school. You are a member of the Student Design Committee, which will be responsible for the layout and design of the park. Use the diagram of the park on page 2 to make your plans as you work with the committee to answer the questions in Part I. Part II will be the final report from the design committee. Your instructor will specify whether this should be a group or an individual report.

Part I: SHOW YOUR WORK.

1. The first step is to construct a fence around the park.

   a. How many yards of fencing will be needed to completely enclose the park?

   b. There should be two gates in the fence to allow people to enter and leave the park. Decide where the two gates should be located.

2. Find the area of the park.

3. A parking lot will be made just outside of the park in the triangular space in the southeast corner. What is the area of the proposed parking lot?

4. The Class of '93 wants to donate a rectangular fish pond for the park. The pond will be made of cement with inside measurements of 8 feet long, 5 feet wide, and 3 feet deep.

   a. What is the volume of the fish pond?

   b. Decide where the fish pond should be located.

5. The Biology Club plans to have a circular flower bed in the park. It will have a diameter of 12 feet. What area will be needed for the flower bed? Use 3.14 for π.

6. A short decorative fence will go around the flower bed.

   a. How much decorative fencing will be needed to go around the flower bed?

   b. Decide where the flower bed should be located.

Part II
Prepare the committee report. Turn in one completed copy of this activity with all of the questions answered and with the appropriate design of the park. Include the following information:

   a. Names of the members of the committee.

   b. Calculations and answers for questions 1 through 6.

   c. Diagram of the park showing the location of the two gates, the parking lot, the fish pond, and the flower bed.

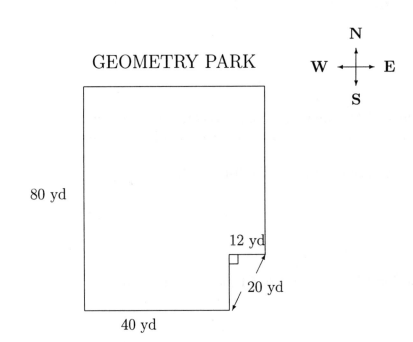

GEOMETRY PARK

Basic Math   Rational Numbers

NOTES TO THE INSTRUCTOR

Summary: Students are asked to use their own words to summarize the rules for operations on rational numbers, and then to use those rules to solve application problems.

Skills Required: Operations with Rational Numbers
Skills Used: Writing Mathematical Procedures, Solving Applications with Rational Numbers

Grouping: Proximity Pairing
Materials Needed and Preparation: Copies of Activity 1-15, one for each student
Student Time: [In Class] 15 to 20 minutes; OPTIONAL: 5 to 10 minutes for Part I only; [Out of Class] none; OPTIONAL: individual homework assignment for Part II

Teaching Tips: Use this activity after covering the operations with rational numbers. Encourage the students to use their own words for the rules in Part I. Part II should be done without a calculator, and students should be reminded to use the correct units in their answers. OPTIONAL: Assign Part II as individual homework.

Grading Tips: Individual Grade; Part I is not graded; 10 points for each question in Part II

Comments: Working with positive and negative numbers can seem confusing for students who have never had algebra. This activity allows them to take control of the rules by writing them in their own words and then to relate the operations to familiar topics such as football, temperatures, and bank accounts.
Connections: Activity 1-3: *Stock Market*
Spin-offs: Allow students to use Part I as a reference during a quiz.

Basic Math   Rational Numbers

Directions: Think about the rules for operations on rational numbers as stated in the text-book. Use the space in Part I to summarize these rules. Use words that will help you solve any problem involving rational numbers. Exchange papers with your partner and compare your statements. Be sure your rules are correct, then use them to answer the questions in Part II.

Part I

In the space below, summarize the rules for working with rational numbers.

ADDITION

SUBTRACTION

MULTIPLICATION OR DIVISION

Part II

1. A student started with a bank balance of $479. Over the next four months, the following changes occurred: −$132,   +$288,   −$116,   +$93
   What was the bank balance at the end of the four months?

2. A person who weighed 132 pounds had the following changes in weight:
        +11 pounds,   −3 pounds,   −9 pounds,   +5 pounds,   −18 pounds
   What was the person's weight after these changes?

3. Use the information in question 2 to find the average change in weight for this person.

4. In a football game, a team had the ball on its 30 yard line. On the next three plays, the team's gain or loss of yardage was as follows: +3,   −7,   +9
    Where was the ball (on what yard line) after the three plays?

5. The temperature in Alaska was −46° at midnight. At noon the next day the temperature was −18°. By how many degrees did the temperature rise?

6. An airplane flew to an altitude of 8,462 feet. Then it recorded the following changes in altitude: −1,907,   −2,312,   +1,050,   +940,   −300
    What was the altitude of the airplane after these changes?

7. In January, the daily low temperatures for one week were as follows:
    −3°,   0°,   5°,   −8°,   −14°,   0°,   6°
    What was the average daily low temperature for that week?

8. Stock in a certain company was $41\frac{1}{8}$ per share. During the week, the market price

    changed as follows: $-\frac{3}{4}$, $+2\frac{1}{2}$, $-4\frac{3}{8}$, $-1\frac{7}{8}$, $+5$

    What was the market price of the stock at the end of the week?

9. The water level in a tank started at 21 feet. The level dropped 3 feet at 10:00 A.M. and then dropped twice that far at noon. Use rational numbers to write a statement that shows the water level at noon.

10. One morning the temperature was 12° below zero. During the rest of the day, the following changes occurred: +6°,   +15°,   −4°,   −9°,   −10°
    What was the temperature after these changes occurred?

NOTES TO THE INSTRUCTOR

Summary: Partners take turns listening and talking as they use critical thinking skills to analyze algebra problems, including practice with finding and correcting errors.

Skills Required: Collecting Like Terms, Evaluating Equations, Solving Equations
Skills Used: Review of skills listed above, Critical Thinking, Making Comparisons, Identifying and Correcting Errors

Grouping: Proximity Pairing
Materials Needed and Preparation: Copies of Activity 1-16, one for each student
Student Time: [**In Class**] 20 to 30 minutes; [**Out of Class**] none

Teaching Tips: Use this activity prior to a test or quiz to review the algebra concepts listed above. Give the students time to read the directions for the activity. You may suggest that they change roles for each problem, or you may keep time and announce, "Switch roles!" every two or three minutes. Monitor the noise level to ensure that "aloud" is not misinterpreted. Problems 3 through 12 require an understanding of how to collect like terms and how to apply the distributive laws.

Grading Tips: Individual or Group Grade; 5 points for each problem

Comments: Encouraging students to talk aloud as they analyze algebra problems gives them an opportunity to use a variety of critical thinking skills. Of particular importance is the ability to communicate thinking processes to other people, such as a partner. Equally important is the ability to listen and follow directions. Being able to find errors is also a valuable skill.
Connections: Activity H: *Learning from Your Mistakes*

Basic Math   Algebra

One way to think through a math problem is to talk to yourself or to a partner. Talking helps you think about what you want to do to solve the problem. In this activity, your partner will listen and help you by writing out your ideas.

Your partner's name: _____

As you work through this activity, you will take turns being either the talking partner or the silent partner.

Talking Partner
THINK and TALK. Say aloud the actions and steps you think are necessary to solve the problem. Focus on thinking and giving directions.

Silent Partner
LISTEN and WRITE. Write down the steps and calculations suggested by your partner. Focus on listening and following directions.

Each time after you switch roles, be sure to check your partner's work on the previous problem. Do you want to make changes or do you agree with your partner's work?

1. Circle the expression that is equivalent to $6x - 5$.
   a. $5x - 6$ b. $-6x + 5$ c. $-5 + 6x$ d. $-5x + 6$

2. Circle the expression that is equivalent to $a - 3b + 2c$.
   a. $2c - 3b - a$ b. $2c + a - 3b$ c. $-3b - 2c + a$ d. $3b + a + 2c$

For problems 3 through 12, collect *like terms* and fill in the box with the missing term. The left side of the equation must be equal to the right side of the equation. Do NOT try to solve.

3. $5 + a + 2a - 4 = 1 + \boxed{}$

4. $a^2 + 8 - 7a^2 + 2 = \boxed{} + 10$

5. $2y + y^2 + \boxed{} + 3y^2 = 9y + 4y^2$

6. $x + 6 + 4x - \boxed{} = 3 + 5x$

7. $a - b - \boxed{} - b = -2b$

8. $11 - 3z + 8z + \boxed{} = 11 + 6z$

9. $3(2 + x) = 6 + \boxed{}$

10. $4(y + 1) - 2y = \boxed{} + 4$

11. $2(x + \boxed{}) = 8x$

12. $\boxed{}(2x + 3y) = 4x + 6y$

89

ERRORS HAVE BEEN MADE in evaluating problems 13 through 16! Circle the error in each problem, and then find the correct value for the expression.

Correct evaluation:

13. Evaluate $a + 4b$ when $a = 6$ and $b = 1$.
$a + 4b = 1 + 4(6) = 1 + 24 = 25$

14. Evaluate $3(x + y)$ for $x = 2$ and $y = 4$.
$3(x + y) = 3(2 + 4) = 6 + 4 = 10$

15. Evaluate $5x - y$ for $x = 3$ and $y = -2$.
$5x - y = 5(3) - 2 = 15 - 2 = 13$

16. Evaluate $\dfrac{8a}{b}$ when $a = 0$ and $b = 4$.
$\dfrac{8(0)}{4} = \dfrac{8}{4} = 2$

ERRORS HAVE BEEN MADE in solving problems 17 and 18! Circle the error in each problem, and then find the correct solution for the equation.

Correct solution:

17. Solve:
$$\begin{aligned} x - 5 &= 15 \\ x - 5 - 5 &= 15 - 5 \\ x &= 10 \end{aligned}$$

18. Solve:
$$\begin{aligned} \frac{1}{8}t &= -8 \\ -\frac{8}{1} \cdot \frac{1}{8}t &= -8 \cdot -\frac{8}{1} \\ t &= 64 \end{aligned}$$

19. Solve and check: $-7 + 2x = -15$ Check:

20. Solve and check: $3.4 - 5x = 6.6$ Check:

NOTES TO THE INSTRUCTOR

Summary: The bragging comments of several friends form the basis of statements to be translated from English into algebra. Student partners also create and solve "algebragging" problems of their own.

Skills Required: Solving Equations
Skills Used: Translating from English to Algebra, Translating into Equations

Grouping: Proximity Pairing
Materials Needed and Preparation: Copies of Activity 1-17, one for each student
Student Time: [**In Class**] 15 to 20 minutes; [**Out of Class**] none

Teaching Tips: Use this activity after a brief introduction to translating from English to algebra. Refer to the example at the beginning of the activity to help students answer questions 1 through 8. Some of the expressions from the first page are used to set up equations on the second page. The last two problems are created by the student partners. Remind students to show an equation as well as a solution for questions 9 through 14.

Grading Tips: Individual Grade; 5 points each for questions 1 through 8; 10 points each for questions 9 through 14

Comments: Translating from English to algebra is a skill that can seem rather far removed from daily life experiences for most students. This activity uses the friendly banter of conversation among peers as a basis for the statements to be translated. It is hoped that this approach will help make the concept seem less threatening and more relevant.
Connections: Activity 3-1: *A Translating Team*

Several friends were talking together one day, and they started "algebragging." Their comments were overheard and recorded in this activity. Work with your partner to translate each comment from English into algebra. When required, set up an equation and solve it. Then try some "algebragging" of your own with your partner!

Questions 1 through 8: Translate each <u>underlined comment</u> into an algebraic expression. Choose a variable to represent the unknown and state what the variable represents.

> For example: My sister is <u>12 years older than you.</u>
>   Let y = you $12 + y$

1. Please! I am <u>twice as tall as Mia</u>!

2. I'll bet Kris only makes <u>$1.00 an hour more than I do.</u>

3. Did you know that Fred lives <u>seven times farther from school than Mia</u>?

4. You have only <u>half of my charm.</u>

5. Wow! George's score was <u>18 points less than Leah's score</u> on the last test.

6. Coach said my jump was <u>3 feet longer than Jerry's jump.</u>

7. Even if we <u>divide our money five ways</u>, we will still have more money than Kris.

8. Your car has maybe <u>a third of the power of my car</u>.

Questions 9 through 12: Set up an equation and solve.

9. Fred lives 14 miles from school (question 3). How far away does Mia live?

10. Remember what I said about George (question 5)? Well, George's score was 74 points. What was Leah's score?

11. Hey, if you doubled Fred's salary and added $2000, you would just equal my salary. I earn $18,000, so how much does Fred earn?

12. Coach told me the total distance Jerry and I jumped (question 6) was 43 feet. Figure out how far Jerry jumped.

IT'S YOUR TURN! Try some "algebragging" with your partner. Create two problems. Translate each problem into an equation and solve.

Partner's name: _____

13.

14.

NOTES TO THE INSTRUCTOR

Summary: Cumulative review problems of topics covered since midterm are arranged so the answer from each problem in a set becomes part of the next problem. Students review together by working problems and checking answers as the problem sets rotate around the group.

Skills Required: Percents, Averages, Order of Operations, Perimeter (including Circumference), Area, Volume (including Cylinder), Pythagorean Theorem, Operations with Real Numbers, Evaluating Expressions, Solving Equations, Translating from English into Algebra and Solving
Skills Used: Cumulative review of the skills listed above

Grouping: Any technique (2 to 5 students per group)
Materials Needed and Preparation: Copies of Activity 1-18, one for each student
Student Time: [**In Class**] 30 to 40 minutes; OPTIONAL: 5 minutes on Day 1, 15 to 20 minutes on Day 2; [**Out of Class**] none; OPTIONAL: individual homework assignment

Teaching Tips: This activity can be used to prepare for the final exam by reviewing the skills and concepts covered since midterm. The activity can be completed with or without a calculator. Before the groups are formed, read aloud the directions on page 1 of the activity. Emphasize the need to check the work carefully on each problem, because the answer from one problem becomes part of the next problem. Distribute the activity sheets so the students can see how various symbols are used to link the answer from one problem to a specific part of the next problem in each set. Remind students to use the correct units, where appropriate, in their answers. OPTIONAL: On Day 1, explain the activity and assign it as individual homework. On Day 2, group students so they can compare answers and discuss questions. Also, for a complete review of basic math skills and concepts, this activity can be combined with Activity 1-10: *Midterm Madness*, which reviews the concepts covered before midterm.

Grading Tips: Participation Grade; OPTIONAL: Group Grade, 5 points for each problem

Comments: Because this activity provides a cumulative review, it encourages students to synthesize skills and concepts as they prepare for the final exam.
Connections: Activity 1-10: *Midterm Madness* reviews topics covered before midterm.

Directions: Each person in the group will start working on a different set of problems. Notice that the answer from each problem should be written in as part of the next problem in the set. Use the lines and symbols as a guide. When you finish one problem, pass your paper to the left so your teammate can do the next problem in the set. When a new problem is passed to you, be sure to check the work of the person ahead of you. As the problems rotate around the group, discuss any errors or questions with your teammates.

Review Set 1

1. Find the missing distance in the figure on the right.

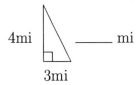

   4mi   _____ mi

   3mi

2. Translate into an equation and solve: _____ students are absent today. This is three more than were absent yesterday. How many were absent yesterday?

   Answer = $\bigcirc$

3. Simplify: $\bigcirc - 5^2 - (-8) = \square$

4. Solve: $3(x - 9) = \square$

   $x = \langle\!\rangle$

5. Find the circumference of a table with a diameter of $\langle\!\rangle$ feet. Use 3.14 for π.

Review Set 2

1. Simplify: $6(-1) + 5 - 3^2 = $ _____

2. Find the average low temperature:
   _____ degrees, 9 degrees, 0 degrees, -3 degrees, 14 degrees

   Average = $\triangle$ _____
                            units?

3. Evaluate: $3a - b$ when $a = \triangle$ and $b = -1$ Answer $= \boxed{}$

4. What is the volume of a box that measures $\boxed{}$ inches long, 4.8 inches wide, and 2.5 inches deep?

         Volume $= \bigcirc \dfrac{}{\text{units?}}$

5. Solve: $8z + 50 = \bigcirc$

Review Set 3

1. What is the area of a circle with a diameter of 20 centimeters? Use 3.14 for π.

         Area $= \triangle\!\!\!\square \dfrac{}{\text{units?}}$

2. $\triangle\!\!\!\square \div (-4) + 90.5 =$ Answer $= \langle\!\!\;\bigcirc\!\!\;\rangle$

3. How much fencing will be needed to go around a garden $\langle\!\!\;\bigcirc\!\!\;\rangle$ feet long and 8 feet wide?

         Answer $= \boxed{}\dfrac{}{\text{units?}}$

4. Evaluate: $5y + \boxed{} - 2x$ when $x = 9$ and $y = -3$

         Answer $= \triangle$

5. What is the volume of the cylinder below if the height is  feet. Use 3.14 for π.

Review Set 4

1. Simplify: $4x - y - x + y = \bigcirc$

2. Solve: $\bigcirc - 15 = 27$ Variable $= \boxed{}$

3. Find the area of a rectangle with a length of 22 meters and a width of $\boxed{}$ meters.

Area $= \langle\bigcirc\rangle \dfrac{}{\text{units?}}$

4. A hamburger has $\langle\bigcirc\rangle$ calories, and 138.6 of those calories comes from fat. What percent of the calories come from fat? Answer $= \underline{}$

5. Translate into an equation and solve: $\underline{}$ of a number is 36. What is the number?

Chapter 2: Introductory Algebra

The activities in this chapter are designed for the course in Introductory Algebra. Some of them may be appropriate for other courses as well. The topic for each activity is listed so the instructor may more readily decide which ones he or she would like to use.

Activity 2-1:	Job Decision	Percents
Activity 2-2:	Grams to Calories	Evaluating Expressions
Activity 2-3:	Please Pass the Equation	Solving Equations
Activity 2-4:	Creating Applied Problems	Problem Solving
Activity 2-5:	The Store Manager's Dilemma	Application Problems with Percent
Activity 2-6:	An Exponential Exploration	Exponents
Activity 2-7:	Get That Factor Off My Back!	Factoring
Activity 2-8:	How High Is That Rocket?	Solving Quadratic Equations
Activity 2-9:	Algebraic Fraction Puzzles	Algebraic Fractions
Activity 2-10:	Graphing Charades	Graphing Vocabulary
Activity 2-11:	The Rent-A-Car Deal	Linear Graphing
Activity 2-12:	Celsius vs Fahrenheit	Graphing Linear Equations
Activity 2-13:	Interpreting Results	Systems of Equations
Activity 2-14:	Rent-A-Car II	Systems of Equations
Activity 2-15:	The Search for the Perfect Square	Radical Expressions
Activity 2-16:	Going a Round with Square Roots	Radical Expressions
Activity 2-17:	The Maximum Playground	Solving and Graphing Quadratic Equations
Activity 2-18:	When Linear Meets Quadratic	Linear and Quadratic Equations

NOTES TO THE INSTRUCTOR

Summary: The student is asked to choose between two jobs where one involves a commission and the other offers an hourly wage and health insurance.

Skills Required: Percents, Averages, Interpreting Data, Commission, Multiplication
Skills Used: Critical Thinking, Percents, Report Writing

Grouping: Count Off, so that the activity can be a mixer for students to meet one another during the first week of class.
Materials Needed and Preparation: Copies of Activity 2-1, one for each student; calculators optional; find cost of health insurance (or assign this to the students)
Student Time: [**In Class**] 20 to 25 minutes; [**Out of Class**] individual homework assignment

Teaching Tips: This activity can be used as a review of percents and basic skills. Hand out a copy of the activity to each student and have him or her do the initial calculations at home as homework. Emphasize that there is no correct job choice. During the next class, have the students work in groups to check their arithmetic on Part 1 and to share their job decisions. Collect the written reports. You may want to allow for resubmissions so that students can learn your expectations.

Grading Tips: Individual Grade; 10 points for each section of Part I; 30 points for Part II; 10 points for report quality. Also, allow students to resubmit if necessary.

Comments: This activity allows students to meet one another and get used to working in groups. The instructor also will have a chance to get to know the students. Student job choices may vary with the economy, individual values, or family structure. Students just out of high school may view this as practice for the future, so their figures may not reflect their present status. Returning adults may view this as a practical activity for their current job situation.

Activity 2-1: Job Decision

You have been offered the following two sales jobs and must choose between them. (*Note*: One job is not necessarily better than the other.) First, answer the questions analyzing the two jobs. Next, use this information to decide which of the jobs you would choose. Last, write a short paragraph explaining why you chose the job you did.

JOB 1: This job pays $15,000 per year plus a commission of 20% of your total annual sales, but no health insurance.

As a skilled job hunter, you ask the first employer what the typical salesperson's annual sales were for the last five years. She gives you the following data sheet:

Sales for Job 1

Year	Sales in dollars
1993	150,000
1992	100,000
1991	170,000
1990	125,000
1989	80,000

JOB 2: This job pays $16.50 per hour with no commission. In addition, 100% of the health insurance premium for your entire family will be paid by your employer.

Part I: The following analysis will help you make your decision about the jobs. Complete each question and then decide which job you would pick.

1. For the five years listed, find the total maximum and minimum pay for Job 1.

2. Find the average sales on Job 1 over the last five years and then find the total average pay for this job.

3. Find the yearly wage earned on Job 2, assuming a 40 hour work week.

4. Estimate what you or your family pays for health insurance for one year. (If you do not have health insurance, your instructor may have hints on how to find some figures.)

5. Find the total compensation for Job 2.

6. Which job would you choose, based on this analysis?

Part II: Write a short paragraph explaining why you chose the job you did.

Writing a Report: Present this activity as a report using your own paper. Organize your calculations from Part I neatly and in a logical order. Write your paragraph from Part II legibly and with complete sentences. Pretend you will be presenting this to your spouse or someone else to explain why you made this job decision.

NOTES TO THE INSTRUCTOR

Summary: Students work together to find the percent of fat in food items. This involves data collection and creating tables as well as evaluating expressions.

Skills Required: Knowledge of Percent
Skills Used: Evaluating Expressions, Collecting and Tabulating Data, Finding Percents

Grouping: Count Off (3 to 5 students per group)
Materials Needed and Preparation: Copies of Activity 2-2, page 1, one for each student, page 2, one for each group; copies of Activity D: *Roles for Groups,* one for each group; copies of Activity A or B, one for each student; plan groups
Student Time: [**In Class**] 15 minutes on Day 1, 15 minutes on Day 2; [**Out of Class**] 15-minute homework assignment

Teaching Tips: This is the first structured group assignment, so it would be a good idea to do Activities A or B before doing Activity 2-2. On Day 1, form groups and use Activity D to help students assign roles. Then have the students choose the food items they want to research; encourage students to choose a variety of foods. Information on fat and calories can be found on most food package labels. Have students predict which items will contain the highest percent of fat. On Day 2, they will bring back their data and complete the tables. If food items contained the information this activity is designed to find, have students check the accuracy of the package's claims.

Grading Tips: Group Grade; 25 points for each step; 25 points for report quality

Comments: The equation of $C = 9G$ can be graphed using the data collected by the students.
Spin-offs: This activity can lead to discussions of healthful eating or researching the fat content of some of the foods available in the school snack bar or cafeteria.

Activity 2-2: Grams to Calories

For health reasons, many people are concerned about the amount of fat in the foods they eat. The fat content is often listed on package labels in grams per serving. To compare the fat content of foods, it is useful to know the percent of calories from fat that a serving of a given food item has. This activity provides a way to convert grams to calories, and then to find the percent of calories from fat in a given food item. The activity is done in four steps: (1) Your group will collect data; (2) you will convert grams of fat to calories; (3) you will find the percent of calories from fat; and (4) you will present your results.

Get Ready

Before you start working, you need to assign roles in your group. You will need the following: A Moderator (to keep everyone on task), a Quality Manager (to make sure your work is top quality), a Recorder (to write everything), and a Messenger (to talk to the instructor). Write your names and roles on the group report.

Solve the Problem

Step 1: Collecting Data

Each member of your group will pick two food items and find the number of grams of fat per serving in each item. Decide which types of food each of you will choose so there are no duplications. You will also find the total number of calories in one serving of that item; this will be used in step 3. Complete the following table:

Name of food	Grams of fat (per serving)	Total calories (per serving)

Step 2: Converting Grams to Calories

This part will be done on the second day. Have the Recorder write the name of each food item and the number of grams of fat (G) for each item in the Conversion Table for Grams to Calories on page 2. The group will then determine the number of calories from fat (C) by evaluating the expression in the last column ($C = 9G$).

Step 3: Finding Percent of Calories

Use the second table on page 3 to find the percent of calories from fat for each item. Be sure to change your decimal or fractional notation to percent notation. Let T = total calories per serving.

Step 4: Presenting the Results

Once you have evaluated all the data, make your own table to present your results. Rearrange the data in descending order of percent of fat from calories. The item with greatest percent of fat from calories will appear at the top.

Write a Report

Write a short paragraph explaining the three tables. Include any comments or analysis your group had about its findings.

<hr>

Group Report

Group Members (with Roles): _____

_____

<hr>

Step 2: Conversion Table for Grams to Calories

Name of food	Grams of fat $= G$	Calories from fat, $C = 9G$

Step 3: Conversion Table for Percent of Calories from Fat

Name of food	Calories from fat $= C$	Total calories $= T$	Percent of calories from fat $= \dfrac{C}{T} \times 100\%$

NOTES TO THE INSTRUCTOR

Summary: Students work together to build basic algebraic equations and take turns solving these equations.

Skills Required: Algebraic Equations
Skills Used: Solving Equations

Grouping: Count Off or Structured Groups, Ability Mixing
Materials Needed and Preparation: Copies of Activity 2-3, page 1, one for each group; page 2, one for each student
Student Time: [**In Class**] 20 to 30 minutes; [**Out of Class**] none

Teaching Tips: Use this activity to replace the lecture on solving basic equations or after the material has been taught to reinforce the concepts. Give students an example on the board or overhead projector before putting them into groups. You may use the example below or make up your own. Go through the instructions on the activity and make sure the students understand what they are to do. Have students do the practice round before asking them to do rounds 1–4 working simultaneously. Each student will record his or her solution on the record sheet, and all will compare solutions with initial equations. The instructor may want to set limits for the numbers used. It is the responsibility of the group to make sure the ending solution matches the original equation.

Grading Tips: Group Participation Grade, or 25 points per problem

Comments: By making their own equations, students gain confidence and find algebra less mystifying.

Example:
First student passes: $\boxed{x = 5}$

Second student passes: $\boxed{2x = 10}$

Third student passes: $\boxed{2x - 7 = 3}$

Fourth student writes on Activity Sheet:

Equation: $\underline{2x - 7 = 3}$

Solution:
$$
\begin{aligned}
2x - 7 + 7 &= 3 + 7 \\
2x &= 10 \\
\frac{2x}{2} &= \frac{10}{2} \\
x &= 5
\end{aligned}
$$

Activity 2-3:   Please Pass the Equation

Group Members: _____

_____

DIRECTIONS: After introducing yourselves, designate which student is to start the first equation and in which direction the equation will be passed.

Practice Round

1. The first student, using his or her own paper, writes "x = <u>some number</u>" and passes this paper to student #2.

2. The second student multiplies both sides of the equation by a number, writes the new equation below the first one, and passes this new equation to student #3.

3. The third student adds or subtracts the same number from both sides of the equation, writes the new equation below the second one, folds over the work done by the second student, then passes this new equation on to student #4.

4. The fourth student writes the final equation on the record sheet and then solves it.

5. The group then checks to see if the solution is the same as the number student #1 started with. If not, the group needs to find out what went wrong.

First Round—*Use only positive integers*
Each student starts an equation on his or her own paper, then passes the paper to the right. All students follow the above directions, working simultaneously, writing and solving equations on their record sheet.

Second Round—*When multiplying, use a negative integer*
Third Round—*When multiplying, use a fraction*
Fourth Round—*When adding or subtracting, use a fraction*

RECORD OF EQUATIONS AND SOLUTIONS

Name:_____

Equation 1:_____

Solution:

Equation 2:_____

Solution:

Equation 3:_____

Solution:

Equation 4:_____

Solution:

NOTES TO THE INSTRUCTOR

Summary: Students work together to create and then solve applied problems that meet the given specifications.

Skills Required: Perimeter, Simple Interest, Sales, Rent-A-Car, and Numbers
Skills Used: Problem Solving; Applications of Perimeter, Simple Interest, Sales, Rent-A-Car, and Numbers; Translating from English into Algebra

Grouping: Structured Groups, Ability Mixing
Materials Needed and Preparation: Copies of Activity 2-4, one for each student
Student Time: [**In Class**] 20 to 30 minutes; [**Out of Class**] group homework assignment

Teaching Tips: Plan this activity for the last part of a class session. This activity should come after students have done the types of problems included. They will need to finish the activity outside of class, so allow some time for students to arrange a group meeting time. On problems 2 and 4, the instructor can assign each group to do all problems, allow groups to choose problem A or B, or assign some groups to problem A and some groups to problem B.

Grading Tips: Group Grade; for four problems, allocate 15 points for each problem (does it make sense, are complete sentences used) and 10 points for each solution.

Comments: Groups approach this activity in different ways. Some groups like to work together to create and solve each problem. Other groups will prefer to divide up the problems. If groups do the latter, the instructor needs to encourage student interaction and to point out that the problems do not require equal amounts of work.
Connections: This same approach can be used for other types of application problems.
Spin-offs: Problems can be used in quizzes or exams. (*Note:* Student problems tend to be more challenging than instructor problems.) A page of the instructor's favorites can also become a good extra credit assignment.

Introduce yourselves. Read through the activity, then decide as a group how you will complete the assignment.

Part I

Each group needs to make application problems using the following specifications:

1. A problem finding the width and length of a rectangle when the perimeter is 360 cm. Give the length in terms of the width. Make sure the width and length are whole numbers.

2. A percent problem:

   A. An interest problem involving a student loan.

   B. A sales problem involving ducks.

3. A rent-a-car problem using two cities in our state.

4. A number problem:

   A. Using the number 36 in the problem.

   B. Using the number 57 in the problem.

Note: Use an equation to solve each problem.

Part II

Now that you have made up the problems, have two different people in your group solve each problem in detail and compare answers. Hand in each problem with its solution. Put the names of those in the group at the top of the page. Make sure to show all of the work that leads to the solution.

NOTES TO THE INSTRUCTOR

Summary: Students are asked to solve a typical problem they would encounter in a retail store that involves percent markup (mark-on) and discounts.

Skills Required: Simple Linear Equations, Percent
Skills Used: Applications of Percent

Grouping: Structured Groups, Ability Mixing
Materials Needed and Preparation: Copies of Activity 2-5, one for each group; copies of Activity D: *Roles for Groups,* one for each student
Student Time: [**In Class**] 20 minutes; [**Out of Class**] group homework assignment

Teaching Tips: Give the students Activity D: *Roles for Groups* to help them work in a structured way in their groups. They may need help coming up with the equation for profit given the wholesale cost and markup: $P = c + mc$, where P = profit, c = wholesale cost, and m = markup in percent. Monitor the work to make sure all students are contributing to the group effort. They should be able to complete problem 1 in class. You may also want to give them in-class help on problem 2. Have students complete the activity as a homework assignment and show you their algebraic expressions for the first two parts before they leave class, as well as their plan for completing the activity. Have them present their answers in a written report.

Grading Tips: Group Grade; 30 points for each problem; 10 for report quality. The extra credit requires several steps and should be given a substantial number of points (e.g., 25 points).

Comments: Some students may not know what a markup (or mark-on) is. Encourage them to ask you for this information. The surprising thing to many students is that just subtracting the percentages does not give the "correct" answer.

Activity 2-5:   The Store Manager's Dilemma

Introductory Algebra   Application Problems with Percent

You are a member of the sales team for a retail clothing store. Yesterday the store received a new shipment of socks from the main warehouse. Your store receives what the parent company sends it. Your manager is upset that the store received these particular items, since they have not been selling well. You and your co-workers decide to advise your manager to put the socks on sale. Before going to her with this proposal, however, you and your co-workers decide to find out what the final profit would be after a sale of 25% off.

Get Ready

Assign roles in your group. You need the following: A Moderator (to keep everyone on task), a Quality Manager (to make sure your work is top quality), a Recorder (to write everything), and a Messenger (to talk to the instructor). List the members of your group and their roles at the top of your report. Arrange a time to meet out of class if necessary. Read the activity completely before you start working.

Solve the Problem

1. The usual markup on socks is 70% of the wholesale cost (this is called a "mark-on" by some accountants). Using algebra, find an expression or equation for the usual retail price using this markup. Let C = wholesale cost and R = retail price. If the wholesale cost for the socks is $2.50, what is the retail price?

2. Using the algebraic expression you found for the retail price, write an equation showing the sale price of the item. Let S be the sale price. If the socks cost $2.50 wholesale, for how much would they be on sale?

3. Simplify your equation from question 2. What is the resulting percent of "markup"? This is how much you will actually make on the socks. Is this what you expected? How much profit will you make on each pair of socks? Does your team think your manager would be happy with this amount of profit?

EXPLORE:

4. Now you and your co-workers want to change the initial markup on the socks in order to increase the profit after the sale. You want to keep the sale at 25% off. What would your initial markup need to be for the resulting markup (as in question 3) to be 45% after the sale? Assume that the wholesale cost to the store is $2.50.

Write a Report

On your own paper, write the answers to these questions as a report. Your instructor will specify whether it should be a group or individual report. Be sure to (1) include your names, (2) show your work, and (3) use complete sentences in answering all questions.

NOTES TO THE INSTRUCTOR

Summary: Working in groups of four, students must work their way through the exponential problems. Each student chooses a type of exponent problem to simplify: multiplying or dividing powers, raising a power to a power, or combining like terms.

Skills Required: Rules of Exponents
Skills Used: Differentiating and Applying Exponent Rules

Grouping: Structured Groups, Ability Mixing or Ability Matching (4 students per group)
Materials Needed and Preparation: Copies of Activity 2-6, one for each group
Student Time: [**In Class**] 30 minutes; [**Out of Class**] none

Teaching Tips: This activity works well as part of a review before an exam on exponents. Form groups, then have students choose skills and record them on the group's activity sheet. Instruct students to record their work and solutions as they solve problems. After they have finished Level 1, have them continue with Level 2, Level 3, and then Level 4, changing skills each time.

Grading Tips: Group Participation Grade when used as a review before an exam. Individual or Group Grade when used as an assignment with each problem in Levels 1 and 2 worth 4 points and each problem in Levels 3 and 4 worth 3 points.

Comments: Many developmental students can learn one technique at a time but have difficulty in combining and differentiating skills. This activity addresses the needs of these students.
Spin-offs: Students who review together in class may decide to study together outside of class.

Activity 2-6:   An Exponential Exploration

Your exploration team must find its way through the following exponential problems. Each one of you will use a special skill to perform a particular operation with exponents.

Directions:

1. Assign skills. Have each member pick a skill and write his or her name next to the skill the person has chosen.

2. Tackle the problems. For each problem identify which skill is required and have the appropriate person work it.

3. Before each level, rotate skills.

Skills:

×   Multiplies powers with like bases   _____

÷   Divides powers with like bases   _____

()   Raises a power to a power   _____

±   Adds or subtracts terms   _____

Problems:

Level 1

1. $y^7 \cdot y^3 \cdot y$

2. $(x^3)^{-4}$

3. $2x^2 - 3x^2$

4. $\dfrac{x^7}{x^8}$

5. $(x^5 y^3)^2$

6. $\dfrac{x^8 y^9}{x^5 y^{11}}$

7. $x^{13} \cdot x^4 y \cdot xy^3$

8. $7x^2 y - 3x^2 y - 4x^2 y$

Level 2 (*rotate skills*)

1. $\dfrac{xy^2 z^3}{xy^5 z^4}$

2. $x^3 - x^4 + x^3$

3. $(x^5 y^4 z)^0$

4. $xyz \cdot x^2 \cdot y^3 \cdot z^4$

5. $3x^2 y - 7xy^2 + 2x^2 y$

6. $\dfrac{2^3 y^0 z^2}{4^3 y^3 z}$

7. $a^3 b^0 \cdot a^4 c^2 \cdot bc$

8. $(2x^3 y^4 z^{10})^3$

Congratulations, you are now ready for Level 3 and 4 problems. Each of the following problems requires more than one skill. As a group, decide which skills are needed and then have the appropriate people work the problems. Be sure to show your work for each problem.

Level 3 (*rotate skills*)

1. $(x^2y)^0 \cdot (x^2y)^3$

2. $3x(x^2 - 4)$

3. $\dfrac{(x^3y)^3}{x^7y}$

4. $(3xy \cdot 4xy^2)^2$

5. $\left(\dfrac{xy^3}{x^5y}\right)^3$

6. $\left(\dfrac{x^2y^3}{xy^5}\right)^0$

Level 4 (*rotate skills*)

1. $(2x^2y)^3 \cdot (xy)^2$

2. $\dfrac{(x^2y^0z^5)^2}{xyz}$

3. $2x^2y(xy^2 - 3xy^2)$

4. $\dfrac{(xyz)^2}{(x^3y^2z)^0}$

5. $\left(\dfrac{x^2y^5}{x^3y}\right)^2$

6. $(2x^2y \cdot 3xy)^2$

NOTES TO THE INSTRUCTOR

Summary: Students match up polynomials in factored and unfactored form. The polynomials are put on the students' backs as they enter class.

Skills Required: Basic Factoring of Polynomials
Skills Used: Factoring and Recognizing Typical Polynomials

Grouping: The whole class
Materials Needed and Preparation: Copies of Activity 2-7 (may want to enlarge when copying), cut out pairs of polynomials; tape or straight pins
Student Time: [**In Class**] 20 minutes; [**Out of Class**] none

Teaching Tips: After factoring has been taught, on one of those gray days in October or March, wake up your students with this activity. As students arrive for class, pin or tape a polynomial on their backs. Make sure to use pairs of polynomials but have them shuffled. If there is an odd number of students, the instructor needs to participate. Students are not allowed to take off the polynomial or sit down until they have been matched with the equivalent form of the polynomial. The whole class must work to find the matching polynomials. When two matching polynomials have been found, have the two students go to the board and check to see whether they match. The student with the polynomial in factored form should multiply his or her problem while the other student factors his or her problem. For a class larger than 28: (1) Put the factors on the backs of 20 to 28 students and have them move around. Or (2) cut and mix the factors and use as a silent matching activity in small groups.

Grading Tips: Individual Participation Grade

Comments: This is one of the few activities that requires a lot of physical activity, which is helpful for students who have a kinesthetic learning style. It can also aid students in learning to factor quickly which helps students when simplifying rational expressions.
Spin-offs: You may want to use some of these polynomials on the next quiz or test in a matching or multiple-choice question.

$a^2 - b^2$	$(a - b)(a + b)$
$x^2 + 4x + 4$	$(x + 2)^2$
$x^2 + 3x + 2$	$(x + 2)(x + 1)$
$2x^2 + 3x - 2$	$(2x - 1)(x + 2)$
$a^2 - 2ab + b^2$	$(a - b)^2$
$x^2 + 4$	Not Factorable
$x^2 - 4$	$(x + 2)(x - 2)$

$a^2 - 3a + ab - 3b$	$(a + b)(a - 3)$
$x^4 - 1$	$(x^2 + 1)(x + 1)(x - 1)$
$x^6 - 2x^5 + 7x^4$	$x^4(x^2 - 2x + 7)$
$a^2 - 10a + 25$	$(a - 5)^2$
$x^2 + 5x + 6$	$(x + 3)(x + 2)$
$x^2 - 5x - 6$	$(x - 6)(x + 1)$
$a^2 + a + 9$	Not Factorable

Activity 2-8: How High Is That Rocket?

NOTES TO THE INSTRUCTOR

Summary: Students use quadratic equations to solve a problem involving the height of a rocket.

Skills Required: Factoring Quadratic Equations
Skills Used: Problem Solving, Solving Quadratic Equations by Factoring, Report Writing

Grouping: Structured Groups
Materials Needed and Preparation: Copies of Activity 2-8, one for each student; copies of Activity D: *Roles for Groups,* one for each student; plan groups
Student Time: [**In Class**] 20 to 30 minutes; [**Out of Class**] group homework assignment

Teaching Tips: It will help your students if you pass out Activity D along with this activity. Divide students into groups then have students assign roles and get started on the activity. The activity requires the missing information that is provided at the bottom of this page. When the messenger from a group comes to the instructor with the correct questions, give her or him the information requested. Make sure that the groups have set a time to meet outside of class before they leave.

Grading Tips: Group Grade; 15 points each for problems 2 through 7; 10 points for report quality. You may want to allow, or require, groups who receive below a 'C' to resubmit their report to improve their grade.

Comments: Some groups may need a bit of help with this activity. The instructor should be ready with questions that will help them discover the answer or question they are looking for. You may want to bring a ball to class to demonstrate the path of a projectile or to draw a diagram of the problem.

Missing Information: The altitude of Rabbit Ears Peak: 8,150 ft.
The altitude of the launch pad: 4,100 ft.

Activity 2-8:   How High Is That Rocket?

Solving Quadratic Equations

While visiting White Sands Missile Range, you observe a rocket fired straight up. Your friend claims the rocket went higher than nearby Rabbit Ears Peak. Curious about this, you ask one of the scientists how high the rocket went. She can tell by looking at you that you have studied algebra and gives you the following equation for the height of a projectile:

$$H = -16t^2 + Vt$$

where H represents the height above the launch pad in feet, t represents the time in seconds, and V represents the initial velocity. She also tells you the initial velocity was 512 ft/sec.

When you return home, you take this problem to your problem-solving team for help in determining if the rocket went higher than Rabbit Ears Peak. The following steps will help you in solving the problem.

Get Ready

First, assign roles in your team. You will need the following: A Moderator (to keep everyone on task), a Quality Manager (to make sure that your work is top quality), a Recorder (to write everything), and a Messenger (to talk to the instructor). List the members of your group and their roles at the top of your report.

Solve the Problem

1. What other information will you need to solve this problem? Formulate your questions and have your messenger ask your instructor.

2. Write the equation in two variables that describes the height of the rocket in relation to time. What type of equation is this?

3. Find the times, in seconds, when the rocket is at a height of 0 ft. Why are there two times? Draw a picture of the rocket's trajectory.

4. Find the times, in seconds, when the rocket is at a height of 3,072 ft. Could this be the highest the rocket went? Why or why not?

5. Considering the information from questions 3 and 4 and your drawing, can you make a guess at the time the rocket achieved its greatest height?

6. Use your estimate to find the greatest height. How can you check to see if this is the greatest height? Check it.

7. Back to the question we started with. Did the rocket go higher than Rabbit Ears Peak? Use the information from questions 1 and 6.

Write a Report

On your own paper, write a report. Be sure to (1) include your names, (2) show your work, and (3) use complete sentences in answering all questions. You may want to add diagrams or pictures to your report.

NOTES TO THE INSTRUCTOR

Summary: Students use puzzle pieces to assemble the correct steps in manipulating algebraic fraction problems. Groups also find a piece that does not belong and give a reason why that piece is incorrect.

Skills Required: Simplifying Algebraic Fractions (Rational Expressions)
Skills Used: Manipulating Algebraic Fractions, Identifying and Understanding Common Errors

Grouping: Structured Groups, Ability Mixing, or Count Off
Materials Needed and Preparation: Copies of Activity 2-9, pages 1 and 2, one for each group; copy and cut out page 3 of algebraic fractions, one for each group. Put all of the pieces for one puzzle in one envelope and number the envelope. Using a new envelope, continue with the next puzzle. When you have done all the puzzles for one group, repeat the process for the remaining groups.
Student Time: [**In Class**] 30 minutes; [**Out of Class**] none

Teaching Tips: This activity can be used after one operation with algebraic fractions has been taught (just copy the appropriate problems), or it can be used at the end of algebraic fractions to help students review. In class, pass out the activity sheet and go over the instructions before forming groups. Give each group a packet of envelopes. Each group will receive one envelope for each puzzle that group will solve. Have the students pass around the envelope for puzzle one, with each student taking a piece of the puzzle until all the pieces are gone. After solving and recording one puzzle, have the groups continue with the next one. The reasons for the incorrect piece will vary and can become a class discussion.

Grading Tips: Group Participation Grade; or 25 points per problem

Comments: This is a good activity to help students sort out all of the rules for simplifying algebraic fractions and build recognition skills. Developmental students often say they know how to do problems yet they continue to do poorly on exams. Many do know "how" to do a process but need practice on "when" to do it. This activity addresses that need.
Connections: Activity H: *Learning from Your Mistakes*

Group Members: _____

_____

Directions:

Your group will be given a number of puzzle envelopes. Each envelope contains pieces to an algebraic fraction puzzle. Appoint a recorder or take turns recording your solutions.

Start with envelope #1; have each member take a piece of the puzzle out of the envelope as you pass the envelope around your group. Keep passing the envelope until all the pieces have been taken. Some people may have more than one piece. As a group, decide which order they go in. *Note:* One piece in each puzzle will not belong there.

Record the order of the puzzle pieces and state why the incorrect piece doesn't work.

PROBLEM 1

Incorrect Piece:

Reason this one doesn't belong:

PROBLEM 2

Incorrect Piece:

Reason this one doesn't belong:

PROBLEM 3

Incorrect Piece:

Reason this one doesn't belong:

PROBLEM 4

Incorrect Piece:

Reason this one doesn't belong:

Instructor: *Read the Notes to the Instructor for Activity 2-9* for suggestions on using this information.

— — — — — — — — — — — — —CUT— — — — — — — — — — — — — — —

1

$$\frac{x^2 + 5x + 6}{x^2 - 9}$$

| 1

$$\frac{(x+3)(x+2)}{(x+3)(x-3)}$$

— — — — — — — — — — — — —CUT— — — — — — — — — — — — — — —

1

$$\frac{x+2}{x-3}$$

| 1

$$\frac{(x+6)(x+1)}{(x-3)(x-3)}$$

— — — — — — — — — — — — —CUT— — — — — — — — — — — — — — —

2

$$\frac{x^2 + 4x + 4}{3x^2 + 6x} \cdot \frac{9x^2}{x^2 - 4}$$

| 2

$$\frac{(x+2)(x+2)}{3x(x+2)} \cdot \frac{3 \cdot 3 \cdot x \cdot x}{(x+2)(x-2)}$$

— — — — — — — — — — — — —CUT— — — — — — — — — — — — — — —

2

$$\frac{(\cancel{x+2})(\cancel{x+2})}{3x(\cancel{x+2})} \cdot \frac{3 \cdot 3 \cdot \cancel{x} \cdot x}{(\cancel{x+2})(x-2)}$$

| 2

$$\frac{3x}{(x-2)}$$

— — — — — — — — — — — — —CUT— — — — — — — — — — — — — — —

2

$$3x(x-2)$$

| 3

$$\frac{a+2}{2} - \frac{a-4}{4}$$

— — — — — — — — — — — — —CUT— — — — — — — — — — — — — — —

3

$$\frac{2(a+2)}{2(2)} - \frac{a-4}{4}$$

| 3

$$\frac{2a + 4 - (a-4)}{4}$$

— — — — — — — — — — — — —CUT— — — — — — — — — — — — — — —

3

$$\frac{2a + 4 - a + 4}{4}$$

| 3

$$\frac{a+8}{4}$$

— — — — — — — — — — — — —CUT— — — — — — — — — — — — — — —

3

$$\frac{2a + 4 - a - 4}{4}$$

| 4

$$\frac{x^2 - 1}{4x + 4} \div \frac{2x^2 - 4x + 2}{8x + 8}$$

— — — — — — — — — — — — —CUT— — — — — — — — — — — — — — —

4

$$\frac{x^2 - 1}{4x + 4} \cdot \frac{8x + 8}{2x^2 - 4x + 2}$$

| 4

$$\frac{(x+1)(x-1)}{4(x+1)} \cdot \frac{8(x+1)}{2(x^2 - 2x + 1)}$$

— — — — — — — — — — — — —CUT— — — — — — — — — — — — — — —

4

$$\frac{(x+1)(x-1)}{4(x+1)} \cdot \frac{8(x+1)}{2(x-1)(x-1)}$$

| 4

$$\frac{(x+1)}{(x-1)}$$

— — — — — — — — — — — — —CUT— — — — — — — — — — — — — — —

4

$$\frac{(\cancel{x+1})(x-1)}{\cancel{2} \cdot 2(x+1)} \div \frac{\cancel{2}(x-1)(x-1)}{8(\cancel{x+1})}$$

|

NOTES TO THE INSTRUCTOR

Summary: Students act out graphing terms by playing charades.

Skills Required: Graphing Terminology
Skills Used: Understanding Graphing Terminology and Concepts

Grouping: Structured Groups, Ability Mixing (3 to 5 students per group)
Materials Needed and Preparation: Copies of Activity 2-10, one for each group; pick terms and write down one or two terms for each group.
Student Time: [**In Class**] 20 to 30 minutes; [**Out of Class**] none

Teaching Tips: Use this activity at least a few days after you have started graphing. The students who see a term acted out will have visual reinforcement of its meaning. Copy the activity before class and write down the term to be used by each group on its activity sheet. Choose terms that have been introduced in class; common graphing terms are listed at the bottom of this page. Go through the directions before forming groups, but aim for a balance between shy and outgoing students. Give the groups 5 minutes to plan what they will do, then have them take turns acting out their terms, taking between 2 to 5 minutes for each charade. Students can choose to portray the actual term or portray syllables that comprise the term.

Grading Tips: Group Participation Grade or Group Grade based on the group's understanding of the term

Comments: This is a good activity for students who need to move. By physically acting out the term, students will become more familiar with the concept it represents. There are many new terms in graphing, possibly more than in other areas of Introductory Algebra. This also is a creative activity.

Graphing Terms:

slope	parallel	axis
y-intercept	perpendicular	ordered pairs
x-intercept	horizontal	quadrants
coordinates	vertical	direct variation
intercepts	origin	inverse variation

Activity 2-10: Graphing Charades

Group Members: _____

_____

Directions: Your group will be given a graphing term to act out in front of the class. You cannot talk while doing your charade. You must get the rest of the class to guess the word while your group acts it out using only *body* language and props. No writing or talking. Your term is given below. You may use this paper to sketch out your ideas. Hand it to your instructor before you take your turn.

Your Term:_____

NOTES TO THE INSTRUCTOR

Summary: Students work together to come up with an equation and graph for rental car agency data. The class as a whole then compares the data using an overhead projector, graphing calculator, or computer and discusses the merits of the different agencies.

Skills Required: Rent-A-Car Applications
Skills Used: Analyzing a Problem Using Graphs

Grouping: Structured Groups, Ability Mixing (3 to 5 students per group)
Materials Needed and Preparation: Copies of Activity 2-11, one for each group; determine which method (projector, graphing calculator, or computer) to use to show all the graphs at once and make necessary preparations; plan Structured Groups.
Student Time: [**In Class**] 30 minutes; [**Out of Class**] none

Teaching Tips: Do this activity after the class has plotted points but before it has graphed lines. Go over the directions and form groups. Each group must assign roles (Moderator, Quality Manager, Recorder, Messenger) and choose a name for its rent-a-car agency. Then the group will send a messenger to the instructor with the assigned roles and name of its agency. The instructor gives the messenger the cost per day and cost per mile for his or her group. Suggested costs are listed below. Make sure that each group has the correct equation before continuing with Part II. Once the groups have finished, the whole class will look at the information. The instructor or a student from each group should write the equation and name of the agency on the board or overhead. Next, all the graphs need to be seen together on the same coordinate plane. One way to do this would be to provide each group with a transparency with the axes already drawn. Have each group graph its equation on the transparency and then overlay the transparencies on the overhead projector. The instructor can also use a graphing calculator or a computer with an overhead projector. Before the class sees the graph of each equation, ask the students to predict where it will start and whether it will be more expensive or less expensive than the other agencies. This process will lead students into a natural comparison of equations and graphs. Discuss the various merits of the different agencies. Concepts such as slope and intercept can be introduced during the discussion.

Grading Tips: Group Participation Grade; or 25 points each for Parts I through IV

Comments: This activity takes less time than it appears it would. It allows an overall introduction to graphing and sets the stage for learning to graph lines.
Connections: Activity 2-4: *Creating Applied Problems* and Activity 2-14: *Rent-A-Car II*

Suggested Costs:	Per Day	Per Mile
1.	$40	$0.40
2.	$50	$0.60
3.	$90	$0.15
4.	$25	$0.50
5.	$75	$0.30

Activity 2-11: The Rent-A-Car Deal

Group Members (with Roles): _____

_____

_____

Have the members in your group choose roles (Moderator, Quality Manager, Recorder, Messenger).

It's spring break and you and your friends have decided to rent a car for a one-day trip. Each group in your class visits a different rent-a-car agency. Make up a name for the agency you visited and send your Messenger to the instructor to find out what your agency charges per day and per mile. Have the Recorder write down this information.

Name of rent-a-car agency:_____

Cost per day:_____

Cost per mile:_____

Part I
Use this information to write an equation that represents the daily cost for renting this car. Let x represent the number of miles traveled and y represent the total cost for the day.

Equation:_____

Part II
Use your equation to fill out the following chart, which shows how much your daily cost will be when you travel 200, 300, or 600 miles in one day.

x = Miles	y = Cost
200	
300	
600	

Part III
Now plot these points on a graph. Have your scale be in 100s of dollars and 100s of miles.

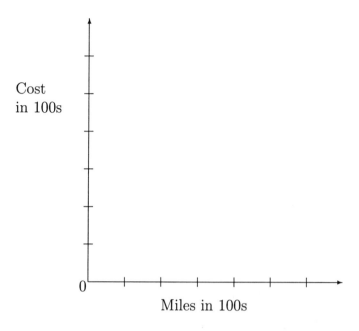

Cost
in 100s

Miles in 100s

Part IV
Answer the following questions:

1. What pattern did your points make?

2. Where do you estimate the point to be that corresponds to 400 miles?

3. Find the cost when $x = 400$, fill out the table, and plot the point on your graph. Was your estimate correct?

4. Connect your points.

5. Where do you estimate your graph will cross the y-axis? How can you find this point using your equation?

6. Use the graph to:

   a. find the cost for 500 miles.

   b. find the miles when the cost is $100.

Part V: Which agency has the best deal?
When all the groups have finished, the Recorder for your group will put your equation on the board or overhead. Your instructor will ask you to put the graph of your equation on an overhead or will use a graphing calculator to show the class all of the lines on one coordinate system. As a class, you will discuss the merits of the different agencies.

NOTES TO THE INSTRUCTOR

Summary: Students are given the freezing and boiling points of water in Celsius and Fahrenheit. They will use this information to find the formula for converting Celsius temperatures to Fahrenheit, to graph the formula, and to write a report.

Skills Required: Finding the Equation of a Line Given Two Points, Graphing Linear Equations

Skills Used: Applications of Linear Equations and Graphing; Connecting Data, Equations, and Graphs; Report Writing

Grouping: Structured Groups (no more than 4 students in a group)

Materials Needed and Preparation: Copies of Activity 2-12, one for each student; copies of Activity D: *Roles for Groups,* one for each student, if needed; plan Structured Groups

Student Time: [**In Class**] 30 minutes; [**Out of Class**] group homework assignment requiring two out-of-class meetings

Teaching Tips: This activity should be used after teaching graphing of linear equations. Allow about 15 minutes at the end of a class session to have groups get started on this activity. As students may still be learning how to work in structured groups, it is recommended that you pass out Activity D when you begin. Pass out Activity 2-12 and go over directions and the time line for completion before you put students in groups. Once in groups, have them complete Activity D, assign roles, and arrange a time to meet. At the next class session have the groups meet for 15 minutes. Answer the messengers' questions and make sure all groups have done Part I correctly. If any group hasn't discovered the idea of writing the information as coordinates, the instructor should aid them by asking appropriate questions. The groups should arrange to meet outside of class one more time before turning in the assignment. You may want to require a rough draft before the final report is due. Specify individual or group reports.

Grading Tips: Individual or Group Grade; 20 points each for Parts I through IV; 20 points for report quality: clarity, completeness, style, and neatness

Comments: This activity meets a number of needs for students who are learning graphing. First, it provides an opportunity to apply algebra to solving a problem. Second, it allows students to make the connection between data and finding an equation. Third, it helps students see information in a different form. The most difficult part of this activity is to go from the given information to seeing it as coordinates. Allowing students to take this home before they start working will give them a chance to "sit in the fog" trying to figure out the problem. This will improve their thinking skills.

Spin-offs: Have students use the equation and graph to convert common temperatures at home or at school.

Activity 2-12:   Celsius vs Fahrenheit

Scenario: To pay off your student loans, you are working as interns at the U.S. Embassy in a foreign country where the temperature is always given in Celsius. Your boss, the U.S. ambassador, knows you have studied algebra and asks you to come up with a linear equation to convert Celsius temperatures to Fahrenheit temperatures. You are given only the following information:

1. Water freezes at 0 degrees Celsius and at 32 degrees Fahrenheit.

2. Water boils at 100 degrees Celsius and at 212 degrees Fahrenheit.

Get Ready: Assign roles in your group. You need the following: a Moderator (to keep everyone on task), a Quality Manager (to make sure that your work is top quality), a Recorder (to write everything), and a Messenger (to talk to the instructor). List the members of your group and their roles at the top of your report. Arrange a time when you can all meet outside of class. Read the activity completely before you start working.

Solve the Problem: When you are working outside of class, have the Messenger write down any questions that you want to ask your instructor.

Part I
Since you neglected to bring along your algebra book, which may have contained the formula, you must use *only* the above information to find a formula for converting Celsius to Fahrenheit. In your equation, use the variables F to represent Fahrenheit and C to represent Celsius. Be sure to show all of your work. *NOTE: You need to show how to get the equation from the given information. Hint:* Think about how you would write the information in properties 1 and 2 above in order to graph it.

Part II
Once you have obtained the formula, use it to complete the chart (round your answers to the nearest tenth, if necessary).

C	F
-50	
	-4
	0
-5	
0	32

Part III
To further impress your boss, the Quality Manager in your group decides to graph your equation. Use the coordinate axes on the next page; notice that the horizontal axis represents Celsius temperature and the vertical axis represents Fahrenheit temperature.

Part IV
Your boss looks at your graph and formula and asks the following questions:

1. What is the slope of your graph?

2. Is this graph increasing (does it have a positive slope)?

3. It looks as though there is one point at which the number of degrees in Celsius is exactly the same as the number in Fahrenheit. a. Find this point on your graph. b. Verify it using your formula.

4. For embassy employees who haven't studied algebra, explain how to use the graph to convert from Celsius to Fahrenheit. For example, given 10° Celsius, how can you use the graph, not the equation, to find the corresponding Fahrenheit temperature?

Write a Report
On your own paper, write a report. Be sure to (1) include your names, (2) show your work, and (3) use complete sentences in answering all questions from Part IV. *Follow any further instructions given by your instructor.*

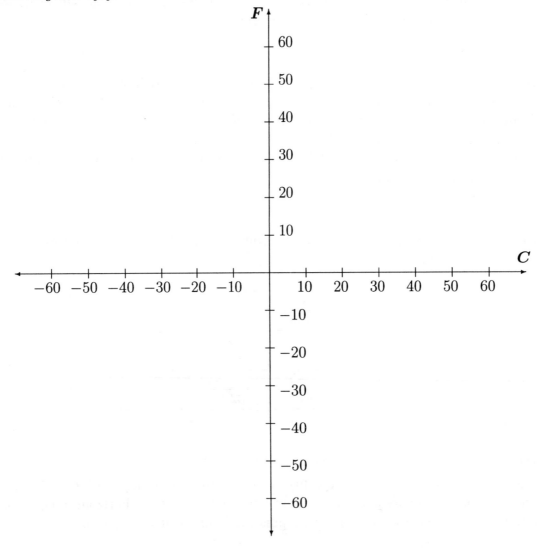

NOTES TO THE INSTRUCTOR

Summary: This activity helps students sort out the different cases where they must interpret results. Cases covered are zero slope, undefined slope, no solution, infinitely many solutions, and the point $(0,0)$. Students are asked to describe the graphical situation that occurs with the given algebraic results.

Skills Required: Finding Slope, Solving Systems of Equations
Skills Used: Analyzing the Exceptional Cases of Slopes and Systems of Equations

Grouping: Structured Groups, Ability Mixing (3 students per group)
Materials Needed and Preparation: Copies of Activity 2-13, one for each student
Student Time: [**In Class**] 20 minutes; [**Out of Class**] none

Teaching Tips: Use this activity after students have learned all methods of solving systems of equations. Go over the directions with students before putting them in groups.

Grading Tips: Group Participation Grade; or 9 points per problem for Parts I and II; 10 points per problem for Part III

Comments: When students confront the situation of interpreting results, they often must think differently from before. Many students have difficulty understanding why undefined slope (no slope) differs from zero slope. Instructors should be prepared to explain why. Recognizing interpretation as a skill that needs to be learned may help students who become frustrated.

Activity 2-13: Interpreting Results

When solving some problems in graphing and systems of equations, there is not always a simple solution. Rather, attempts at solving a problem may give results that must be interpreted. Interpreting results may be a new and challenging part of math for you. As calculators and computers are more and more in use, interpreting results becomes an important math and job skill. The following activity provides more experience with the different results that may arise in this course.

Before you start working, assign the following roles in your group: Recorder (to write everything), Grapher (to do the graphs), and Solver (to work the problems). Write your names and roles below.

Group Members (with Roles): _____

_____

_____

Part I: Slopes of Lines
For each set of points below, complete the following steps:

   a. Find the slope of the line.

   b. Graph the line using the given points for each line.

   c. Choose one of the following to describe your interpretation.

       i. slope $= 0$
      ii. not defined
     iii. slope is a number other than 0 (write the number)

1. $(2, 1)$ and $(0, 1)$

2. $(3, 2)$ and $(3, -2)$

3. $(-1, 5)$ and $(3, -2)$

4. $(1, 4)$ and $(3, 4)$

5. $(-1, 0)$ and $(-1, 3)$

Part II: Systems of Equations (*Switch Roles*)
For each system of equations on page 2, complete the following steps:

   a. Solve the system by the addition method.

   b. Graph the two equations.

   c. Choose one of the following interpretations of your results.

       i. no solution
      ii. infinitely many solutions
     iii. one point (write the coordinates)

1. $2x + 3y = 12$
   $-2x - 3y = 6$

2. $x + y = 2$
   $-x - y = -2$

3. $2x - 3y = 0$
   $x + y = 0$

4. $x + y = 1$
   $2x + 2y = 1$

Part III

Use your results to fill in the rest of the chart below by writing the following statements in the correct location:

not defined   slope $= 0$ ~~vertical line~~ $0 = 3$

parallel lines   ~~slope $= \dfrac{0}{3}$~~   slope $= \dfrac{5}{0}$ ~~no solution~~

~~equations represent the same line~~ $0 = 0$ horizontal line $x = 0$ and $y = 0$

infinitely many solutions $(0,0)$ the origin

	Results	Graphical situation	Your interpretation
S l o p e s	slope $= \dfrac{0}{3}$		
		vertical line	
		equations represent the same line	
S y s t e m s			no solution

NOTES TO THE INSTRUCTOR

Summary: Students are asked to analyze two rental car agency costs. The analysis involves writing equations, graphing linear equations, solving a system of linear equations, and using a graph to analyze a situation.

Skills Required: Solving Systems of Equations
Skills Used: Connecting Equations and Graphs, Reviewing Linear Equations

Grouping: Structured Groups
Materials Needed and Preparation: Copies of Activity 2-14, one for each student
Student Time: [**In Class**] 15 to 20 minutes; [**Out of Class**] time to meet once between class sessions

Teaching Tips: Use this activity after material on solving systems of equations has been taught. Introduce and go over directions at the end of a class session. Have students meet in their groups to get started and to arrange a time to meet outside of class. Announce whether students will turn in a group or individual report. For question 8, assume there is no additional charge for one-way rental.

Grading Tips: Group Grade; 10 points for each problem; 20 points for report quality

Comments: As the semester nears the end, activities such as this can help students make the connections between the different areas they have studied. Seeing how the material is interconnected can increase the students' retention abilities. This activity extends the work begun in Activity 2-11: *The Rent-A-Car Deal.*
Connections: Activity 2-11: *The Rent-A-Car Deal*

Activity 2-14: Rent-A-Car II

Scenario: As Thanksgiving approaches, you decide to accept your aunt's invitation to visit. Since your car isn't currently reliable, you decide to rent a car for the trip to your aunt's house. You have narrowed down your choices to two companies. Company A charges $75 per day and $0.30 per mile. Company B charges $25 per day and $0.60 per mile. You decide to analyze these costs in order to decide which company to use.

Get Ready: Assign roles in your group. You will need the following: a Moderator (to keep everyone on task), a Quality Manager (to make sure that your work is top quality), a Recorder (to write everything), and a Messenger (to talk to the instructor). List the members of your group and their roles at the top of your report. Arrange a time when you can all meet. Read all of the activity before you start working.

Solve the Problem:

1. As a first step in your analysis, write the two equations that represent the daily cost for a car from Company A and a car from Company B. Let x represent miles and y represent the cost.

2. Next, graph both equations on the same axes. Determine your scale. (Should each mark on the axis stand for 1, 10, 50, or 100 miles or dollars?) You can use different scales for the x- and y-axes. After you graph each equation, be sure to label it.

Use your graph to answer the next two questions.

3. If you rent the car for one day and drive 100 miles, which car is a better deal?

4. If you rent the car for one day and drive 300 miles, which car is a better deal?

5. On your graph, your two lines should intersect in a point. Find this point algebraically. What is the cost at this point? What is the number of miles? What are the different ways you can express this particular number? Which of these expressions is preferred by your instructor?

6. When is Company A a better deal? When is Company B a better deal? Write your answers as inequalities.

7. You find out it is 250 miles to your aunt's house. If you drive round trip and have the car from Wednesday at 4 P.M. until you get back Sunday before 4 P.M., which car will be the better deal? Assume you do not drive while at your aunt's house. Show your work and how much each will cost.

8. There is a possibility that you could get a ride home on Sunday with your uncle, who will be going past your college. If you drive only one way and drop the car off at the rental agency a block from your aunt's house, will your choice of companies change?

Write a Report

Using your own paper, write a report. Your instructor will specify whether this should be a group or individual report. Be sure to (1) include your names, (2) show your work, and (3) use complete sentences in answering all questions.

Activity 2-15: The Search for the Perfect Square

NOTES TO THE INSTRUCTOR

Summary: Students find the perfect squares between 1 and 100. After identifying them, the students use them to find square roots.

Skills Required: Squaring Whole Numbers
Skills Used: Finding Perfect Squares, Square Roots, Simplifying Square Roots

Grouping: Proximity Pairing
Materials Needed and Preparation: Copies of Activity 2-15, one for each student
Student Time: [**In Class**] 20 to 30 minutes; [**Out of Class**] none

Teaching Tips: Use this activity at the beginning of the chapter on radical notation. It can replace lecture on numerical square roots. Pair students up and pass out the activity. Make sure students find all the perfect squares in Part I before continuing. Have each student do the work on his or her own paper as this can be used as a reference sheet for the rest of their work with square roots. When students are finished, have the class discuss any patterns they may have seen in doing the activity and which numbers have no radical sign in the answer. Use of calculators is appropriate.

Grading Tips: Group Grade; 30 points for Part I; 40 points for Part II; 30 points for Part III

Comments: Many algebra students do not have a good understanding of square roots. This activity provides further practice in this area as well as teaches students to identify numbers with perfect square factors.

Many composite or nonprime numbers contain hidden perfect squares. Finding these perfect square factors of a number allows us to simplify radical expressions.

$$\textbf{Example: } \sqrt{18} = \sqrt{9} \cdot \sqrt{2} = 3\sqrt{2}$$

This activity will give you practice finding perfect squares and simplifying radical expressions.

Part I: Together

Consider the integers from 2 through 100. Find all the perfect squares and their positive square roots.

1. List all the perfect squares:
   Example: 4

2. Take the positive square root of each perfect square.
   Example: $\sqrt{4} = 2$

Part II: Individually

1. Student #1 lists all the odd perfect squares while student #2 lists all the even perfect squares after 4.

2. Each student then writes 3 multiples of each perfect square on his or her table.

3. Switch papers and simplify the square root of each multiple.

4 is done as an example

Perfect Squares	Multiples of Perfect Squares	Square Root of Multiples
4	12	$\sqrt{12} = \sqrt{4} \cdot \sqrt{3} = 2\sqrt{3}$
	20	$\sqrt{20} = \sqrt{4} \cdot \sqrt{5} = 2\sqrt{5}$
	36	$\sqrt{36} = \sqrt{4} \cdot \sqrt{9} = 6$

4. Switch papers again and check each other's work.

Part III: Together Again

Now that you have had some practice at finding perfect squares, work together to simplify the following radical expressions.

1. $\sqrt{32}$

2. $\sqrt{200}$

3. $\sqrt{98}$

4. $\sqrt{125}$

5. $\sqrt{162}$

6. $\sqrt{64}$

7. $\sqrt{72}$

8. $\sqrt{27}$

NOTES TO THE INSTRUCTOR

Summary: Students take turns manipulating radical expressions. Four types of operations are included: multiplication, division, addition/subtraction, and rationalization of denominators.

Skills Required: Manipulating Radical Expressions
Skills Used: Reviewing and Identifying Different Radical Expressions

Grouping: Structured Groups, Ability Mixing (3 to 5 students per group)
Materials Needed and Preparation: Copies of Activity 2-16, one per student; plan groups
Student Time: [**In Class**] 20 minutes; [**Out of Class**] none

Teaching Tips: Use this activity after students have studied manipulating radical expressions. This activity can be used as a review before an exam. Pass out the activity and go over the instructions with the students. If there are groups of 3, have them do three rounds; groups of 4, four rounds; groups of 5, five rounds. The rounds increase in difficulty, so instructors may want to specify which rounds each group should do.

Grading Tips: Group Grade; 25 points per problem, then divide by the number of rounds

Comments: Developmental students need practice seeing problems of different types on the same page. Recognizing the type of problem and the approach needed may often be more difficult than the process of solving it.
Connections: Students may use the results of Activity 2-15: *The Search for the Perfect Square* while doing this activity.

In this activity you will be taking turns simplifying different types of radical expressions. The activity is set up in three to five rounds. Your instructor will tell you how many rounds your group should do. Each round has four problems, one each of multiplication, division, addition, and rationalization of the denominator. Decide which student will begin each round. Each student does problem 1 of the round he or she is starting, then passes the paper to the student to the right, who will then check the work and begin the second problem. Continue to pass the problems until all problems have been done. When you receive a new paper, check all the problems already done before doing your problem. After the last problem has been done, pass the papers one more time to check answers before handing your papers in.

Group Members: _____

_____

Round 1

1. $\sqrt{2} \cdot \sqrt{18}$

2. $\dfrac{\sqrt{18}}{\sqrt{2}}$

3. $\sqrt{18} + \sqrt{2}$

4. $\dfrac{2}{\sqrt{18}}$

Round 2

1. $\sqrt{x^3} \cdot \sqrt{5xy^5}$

2. $\sqrt{\dfrac{25}{a^2}}$

3. $\sqrt{81x^3} - \sqrt{4x}$

4. $\dfrac{3}{\sqrt{x}}$

Round 3

1. $\sqrt{5ab^2} \cdot \sqrt{15a^2b}$

2. $\dfrac{\sqrt{63y^3}}{\sqrt{7y}}$

3. $3\sqrt{48} - 2\sqrt{27} - 3\sqrt{12}$

4. $\dfrac{5}{\sqrt{5} - 2}$

Round 4

1. $\sqrt{6x^9y^4} \cdot \sqrt{12x}$

2. $\dfrac{\sqrt{54x^7y^2}}{\sqrt{16y^2}}$

3. $2a^2\sqrt{12a^2} - \sqrt{3a^6} + \sqrt{9a^9}$

4. $\dfrac{\sqrt{12}}{\sqrt{8a}}$

Round 5

1. $(\sqrt{a} - \sqrt{2})(\sqrt{a} + \sqrt{8})$

2. $-\sqrt{\dfrac{98}{2}}$

3. $\sqrt{4x + 8} - \sqrt{x + 2}$

4. $\dfrac{2 - \sqrt{5}}{4 - \sqrt{2}}$

NOTES TO THE INSTRUCTOR

Summary: Students solve a maximization problem by graphing a quadratic equation and finding the vertex of the parabola.

Skills Required: Solving and Graphing Quadratics
Skills Used: Concept of Maximum, Application of Quadratics, Report Writing

Grouping: Structured Groups, Ability Mixing (3 to 4 students per group)
Materials Needed and Preparation: Copies of Activity 2-17, one for each group; plan Structured Groups
Student Time: [**In Class**] 10 minutes; [**Out of Class**] time for groups to meet once or twice

Teaching Tips: Once graphing parabolas has been introduced, students can do this activity. Hand out Activity 2-17 at the end of a class; go through the directions with the students and set a due date. The groups should be of mixed ability. Specify your requirements for the report. You can choose to present the information on page 2 yourself or leave it to the groups.

Grading Tips: Group Grade; 20 points each for problems 1–4; 20 points for report quality and extra effort

Comments: This activity can be done earlier in the course if you teach graphing after solving quadratic equations by factoring. The class can also discuss what is happening at the points where the graph crosses the x-axis. This activity may be a stretch for your students. If so, you may want to use it as extra credit.

Spin-offs: If your school has a day-care center, have students find the playground area and see whether it is the maximum area for the amount of fencing.

Activity 2-17: The Maximum Playground

Introductory Algebra   Solving and Graphing Quadratic Equations

Scenario: One of the local hardware stores has donated 60 feet of fencing to enclose a small play yard for children at your college day-care center. Your group has volunteered to put up the fence. The day-care director is trying to figure out how to enclose the maximum area with the 60 feet of fence. Since you have studied algebra, your team shows her how to solve the problem.

Get Ready: Assign roles: Moderator, Quality Manager, Recorder, and Messenger. Read through the activity before you begin to work. There is important information on page 2 to help in solving this problem. Write down any instructions given by your instructor.

Solve the Problem:

1. The director has decided that she can get the largest playground by using one wall of her building for one side of the playground. She also wants it to be a rectangle. Draw a picture of the playground and be sure to include the building.

2. Make a quadratic equation for the area in terms of one of the sides (call these length and width) by following these steps:

   a. Define variables.

   b. Write an equation that relates the length and width to the 60 feet (look at your picture).

   c. Write the formula for the area of a rectangle.

   d. Using substitution, combine these two equations to get one quadratic equation.

3. Graph the equation.

4. Answer these questions leading to the solution.

   a. What point on your graph corresponds to the maximum area?

   b. What is the maximum area?

   c. What dimensions will you use for the playground?

Write a Report: Using your own paper or computer, write a group report. Be sure to show all your work and use complete sentences. List your names and your roles.

Further Information: This is a maximization problem. Many problems in math involve finding a maximum or a minimum value. In calculus, students learn how to do this by taking the derivative. However, there is a way to find a maximum or minimum value of a quadratic equation from the graph of the equation. This activity uses the graph to find a maximum value. Look at the following two graphs.

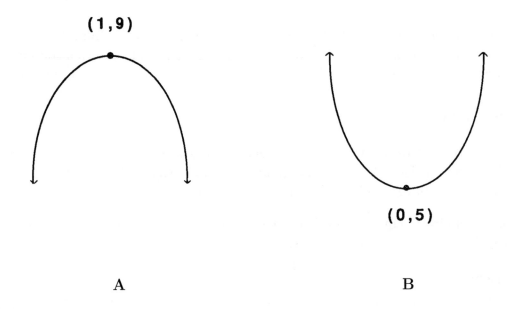

A   B

Which graph has a maximum point? What is this point called? What is the value of y at this point?

NOTES TO THE INSTRUCTOR

Summary: Students in pairs create and graph linear and quadratic equations.

Skills Required: Solving and Graphing Linear and Quadratic Equations, Solving Systems of Linear Equations

Skills Used: Critical Thinking, Reviewing Linear and Quadratic Equations

Grouping: Ability Matching (2 students per group)

Materials Needed and Preparation: Copies of Activity 2-18, one for each student; plan groups

Student Time: [**In Class**] total of 25 minutes over two different class times; [**Out of Class**] individual and group homework assignment

Teaching Tips: This is an end-of-course activity that helps students to review and to connect many of the concepts covered in Introductory Algebra. Use it after all material has been covered or while finishing the chapter on quadratic equations. Do a careful pairing of students of matched ability. Give out the activity at the end of a class session; allow about 10 minutes. Put students in pairs and have them do Part I for homework by the next class session. During the next session, spend 15 minutes having them check each other's work and do Part II. As this is Ability Matching, some pairs will need assistance. Ask students to meet outside of class to finish this assignment and to write a report for the next class.

Grading Tips: Group Grade; 6 points for each problem; 22 points for report quality

Comments: Top students can be challenged by telling them to make their equations as simple as possible. Once they realize they can do this, this activity does not take very long. It is also faster to grade than it appears. Graphing calculators or computers could be used to check answers.

Activity 2-18: When Linear Meets Quadratic

During this course, we have been studying linear and quadratic equations. For this activity, you and your partner will write linear equations and a quadratic equation. Each of you will write linear equations and do Part I, and then work together to do Part II. Write a report on your own paper showing all work and indicating who did each part.

Part I: The Linear Equations

1. Each student writes a linear equation in two variables. Use x and y.

2. What is the slope of your line?

3. a. What is the y-intercept for each?

   b. What is the x-intercept for each?

4. Give the coordinates of two other points on each line.

5. Graph and label both equations on the same axes. Label the y-intercept and two other points.

6. a. Do your two graphs intersect?

   b. Solve algebraically for the coordinates of the point of intersection. Do this even if your graphs do not intersect. How will your algebra results indicate that there is no intersection point?

Part II: The Quadratic Equation (do together)

1. Write a quadratic equation in the variables x and y.

2. Will your graph be concave up or concave down? Why?

3. a. List the terms of your equation.

   b. List the coefficients of each term.

   c. What is the constant term?

4. What is the y-intercept?

5. What are the x-intercepts (if they exist)? Which method did you use to find them?

6. What are the coordinates of the vertex of this parabola?

7. Graph your quadratic equation. Label the vertex, the x-intercepts, and two other points.

Cross-reference of Algebra Activities

To assist instructors in providing their students with a variety of algebra activities, the following cross-reference lists Intermediate Algebra activities that may be used to supplement Introductory Algebra activities. Also listed are Introductory Algebra activities that may be used to supplement Intermediate Algebra activities.

Introductory Algebra activities	Supplement with
Activity 2-1: Job Decision	Activity 3-2: OOOP! Order of Operations Game
Activity 2-2: Grams to Calories	Activity 3-1: A Translating Team
Activity 2-5: The Store Manager's Dilemma	Activity 3-6: Orange Juice Demonstration
Activity 2-8: How High Is That Rocket?	Activity 3-8: Building a Sunroom
Activity 2-9: Algebraic Fraction Puzzles	Activity 3-9: Making a Bid
	Activity 3-10: Water Works!
Activity 2-11: The Rent-A-Car Deal	Activity 3-4: Building a Road
Activity 2-12: Celsius vs Fahrenheit	Activity 3-5: Parallel and Perpendicular Explorations
Activity 2-13: Interpreting Results	Activity 3-7: Coffee on the Run!
Activity 2-17: The Maximum Playground	Activity 3-12: Not That Sunroom Again!

Intermediate Algebra activities	Supplement with
Activity 3-1: A Translating Team	Activity 2-2: Grams to Calories
Activity 3-2: OOOP! Order of Operations Game	Activity 2-6: An Exponential Exploration
	Activity 2-7: Get That Factor Off My Back!
Activity 3-3: Working with Sets	Activity 2-1: Job Decision
	Activity 2-3: Please Pass the Equation
Activity 3-4: Building a Road	Activity 2-10: Graphing Charades
Activity 3-5: Parallel and Perpendicular Explorations	Activity 2-11: The Rent-A-Car Deal
	Activity 2-12: Celsius vs Fahrenheit
Activity 3-6: Orange Juice Demonstration	Activity 2-4: Creating Applied Problems
	Activity 2-5: The Store Manager's Dilemma
Activity 3-7: Coffee on the Run!	Activity 2-13: Interpreting Results
	Activity 2-14: Rent-A-Car II
Activity 3-8: Building a Sunroom	Activity 2-8: How High Is That Rocket?
Activity 3-9: Making a Bid	Activity 2-9: Algebraic Fraction Puzzles
Activity 3-12: Not That Sunroom Again!	Activity 2-17: The Maximum Playground
Activity 3-13: Getting Rational About Inequalities	Activity 2-15: The Search for the Perfect Square
	Activity 2-16: Going a Round with Square Roots

Chapter 3: Intermediate Algebra

The activities in this chapter are designed for the course in Intermediate Algebra. Some of them may be appropriate for other courses as well. The topic for each activity is listed so the instructor may more readily decide which ones he or she would like to use.

Activity 3-1:   A Translating Team   Translate from English into Algebra

Activity 3-2:   OOOP! Order of Operations   Order of Operations
                 Game

Activity 3-3:   Working with Sets   Set Union and Intersection

Activity 3-4:   Building a Road   Slope (Grade)

Activity 3-5:   Parallel and Perpendicular   Parallel and Perpendicular Lines
                 Explorations

Activity 3-6:   Orange Juice Demonstration   Application Problems Involving Percent
                                                             Solution

Activity 3-7:   Coffee on the Run!   Applications of Systems of Equations
                                                             and Inequalities

Activity 3-8:   Building a Sunroom   Solving Quadratic Equations by
                                                             Factoring

Activity 3-9:   Making a Bid   Solving Work Problems

Activity 3-10:   Water Works!   Ratio and Proportion

Activity 3-11:   Complex Numbers   Complex Numbers

Activity 3-12:   Not That Sunroom Again!   Solving Quadratic Equations by the
                                                             Quadratic Formula

Activity 3-13:   Getting Rational About   Quadratic and Rational Inequalities
                 Inequalities

Activity 3-14:   A Carnival of Conics—   Conic Sections—Parabolas
                 Parabolas

Activity 3-15:   A Carnival of Conics—Circles   Conic Sections—Circles and Ellipses
                 and Ellipses

Activity 3-16:   A Carnival of Conics—Circles,   Conic Sections—Circles, Ellipses, and
                 Ellipses, and Hyperbolas   Hyperbolas

Activity 3-17:   Functioning with Spreadsheets   Functions

Activity 3-18:   How Much Space Do We Need?   Exponential Growth

NOTES TO THE INSTRUCTOR

Summary: Students are given the task of translating English phrases and sentences into algebra and vice versa. The task is presented in the scenario of a translation team on an intergalactic ambassadorial mission.

Skills Required: Knowledge of Variables, Basic Mathematics
Skills Used: Translating from English to Algebra, Recognition of Key Phrases

Grouping: Proximity Pairing or Count Off
Materials Needed and Preparation: Copies of Activity 3-1, one for each group
Student Time: [**In Class**] 20 to 30 minutes; [**Out of Class**] none

Teaching Tips: This activity should be a review with the focus on students' recognizing that there could be extraneous English words that are not translatable into mathematics and picking out the key words or phrases that *are* translatable. Emphasize that they will need to use different variables for each sentence. This is a good time to help them learn how to work effectively in groups (see Activity C: *To the Student*). Structure the group interactions so that you don't have two or three students working individually even though they are supposedly in a group. In Part II, students are to come up with English sentences to match mathematical expressions. This section can be given as extra credit or as a test to see whether students have mastered the concept.

Grading Tips: Group Grade; 10 points each for the seven sentences in Part I; 10 points each for the three expressions in Part II

Connections: Concept may be used with some simplification for Basic Math or Introductory Algebra.
Spin-offs: Share the sentences the students translate from Part II in the school newspaper.

It is the year 2510 and you are on an ambassadorial mission to the planet Mathematix. This planet is so named because its inhabitants speak almost entirely in mathematical symbols. You are a new cadet assigned to the Translation Team.

Part I

Below is a sample transmission to be sent to Mathematix. You and a team of other cadets may begin showing your translation abilities by translating the English phrases and sentences to the appropriate algebraic expressions. In each sentence you will need to determine which portion is translatable. Be sure to label variables (use a different set of variables for each sentence) and state your assumptions. The first sentence is done for you, as an example.

"We need to add two more members to one of our exploration teams. They have found fifty percent of the plants you listed. The costs have been reduced by thirty percent of the original estimate. We divided the sectors equally among the four exploration teams. The difference between the number of insects predicted and those observed was ten. A third of the supplies you gave us are still unopened. The number of animals in sector four was five more than the predicted three times those in sector three. Thanks for all of the help."

Example: T: Number of people on the exploration team; $T + 2$

Part II

You have succeeded admirably in the task, and now your supervisor has given you one more task to complete. Mathematix has just sent a transmission for your team to translate. Give it your best shot!

F: Number of furry creatures found; $F + 7$.
S: Supplies needed; $S \times 30\%$.
W: What is wanted; $\frac{1}{2}(W)$.

NOTES TO THE INSTRUCTOR

Summary: This is a game for three to five players that allows students to practice performing operations in the correct order. Play is determined by the accepted order of operations.

Skills Required: Mathematical Operations
Skills Used: Order of Operations

Grouping: Proximity Pairing or Count Off (3 to 5 students per group)
Materials Needed and Preparation: Copies of Activity 3-2, page 1, one for each group; page 2, one for each problem; pick problems from a textbook, fill in the first box on the page 2 template, make copies for each group as needed (see Teaching Tips)
Student Time: [**In Class**] 15 to 30 minutes; [**Out of Class**] none

Teaching Tips: This activity can be used to teach order of operations. Demonstrate how to play the game to the entire class before forming groups so that they all understand the rules. An example appears on page 2 of the Notes to the Instructor. Begin by giving each group a different problem and then having them present their work and solutions on the board when finished. Monitor the groups to make sure everyone is involved in the process. One way to check for this is to scan the room and make sure the game templates are being passed around. If a template is stalled with one group member, check to see whether the group needs help or one person is filling in the entire template. After the groups show some ability, give each group the same problem and have them race to see which team gets the correct answer first. OPTIONAL: Rather than writing the problems on the game templates, list problems on the board or overhead projector and give each group enough blank templates, or ask them to use their own paper, to complete all of the problems.

Grading Tips: Participation Grade; OPTIONAL: Group or Combined Grade; 5 points for knowing where to start in the problem based on parentheses and 5 points for each correct operation completed

Comments: This activity may be used at any level, depending on the type of problem selected.

NOTES TO THE INSTRUCTOR

Example 1:
Simplify: $5\{-2 + [4 - 2(5-3)^2] \div 4\}$

Game Template

Original Expression $\boxed{5\{-2 + [4 - 2(5-3)^2] \div 4\}}$

A	$5\{-2 + [4 - 2(2)^2] \div 4\}$
E	$5\{-2 + [4 - 2(4)] \div 4\}$
M	$5\{-2 + [4 - 8] \div 4\}$
A	$5\{-2 + [-4] \div 4\}$
M	$5\{-2 + (-1)\}$
A	$5\{-3\}$
M	-15

The answer should now be in the last box of the game template.

Example 2:
Simplify: $3\{[6(x-4) + 5^2] - 2[5(x+8) - 10^2]\}$

Game Template

Original Expression $\boxed{3\{[6(x-4) + 5^2] - 2[5(x+8) - 10^2]\}}$

E	$3\{[6(x-4) + 25] - 2[5(x+8) - 100]\}$
M	$3\{[6x - 24 + 25] - 2[5x + 40 - 100]\}$
A	$3\{[6x + 1] - 2[5x - 60]\}$
M	$3\{6x + 1 - 10x + 120\}$
A	$3\{-4x + 121\}$
M	$-12x + 363$

1. Before Play Begins

You will need from three to five players on a team. Make sure an expression is entered in the first box on the game template. Assign each player a role or operation: **E**—Exponents, **M**—Multiply/Divide (may have two players working as a team, one who multiplies and one who divides); **A**—Add/Subtract (may have two players working as a team, one who adds and one who subtracts).

2. Determining a Player's Turn at Play

All players will analyze the expression together and decide which part to complete. If there are parentheses, then the players will decide whether what is inside the parentheses needs simplification. **E** *will perform his or her operation before* **M**, *and* **M** *will perform his or her operation before* **A**. Each player's turn ends when he or she encounters an operation that precedes his or hers.

3. Each Player's Task

E—Exponents

> Working left to right, **E** raises arguments to the indicated power. If there are polynomials, then **E** will raise the polynomial to a power (e.g., squaring a binomial).

M—Multiply/Divide

> Working left to right, **M** performs multiplications and divisions in order. If there are polynomials, **M** will use the distributive property or FOIL. Multiplication and division may be done by two players working as a team. If these tasks are split, then the two players will perform their assigned operations in order, taking turns as needed.

A—Add/Subtract

> Working left to right, **A** performs additions and subtractions in order. If there are polynomials, **A** will combine like terms. Addition and subtraction may be done by two players working as a team. If these tasks are split, then the two players will perform their assigned operations in order, taking turns as needed.

4. Recording Results

Results are recorded on the game template with each player rewriting the new form of the expression in the next box and initialing his or her work with **E**, **M**, or **A**.

Game Template

Original Expression

E, M, or A

NOTES TO THE INSTRUCTOR

Summary: Students are asked to make two sets of kitchen, car, or outside items. In class, they find the unions and intersections of all the sets in their group.

Skills Required: How to Write Sets
Skills Used: Set Union, Set Intersection

Grouping: Proximity Pairing or Count Off (2 to 3 students per group)
Materials Needed and Preparation: Copies of Activity 3-3, one for each student; OP-TIONAL: Make sets for the students and give one copy of the activity to each group.
Student Time: [**In Class**] 30 minutes; [**Out of Class**] 30 minutes, individual assignment; OPTIONAL: none

Teaching Tips: Hand out the activity the class period before groups will work on it. Assign Part I as homework. The next class period, discuss briefly how to write a set using roster notation and/or set-builder notation. Make sure the students understand the activity before they form groups. Have the students in each group label the sets in such a way that each set has a unique label. You may use letters A, B, C, and so on, or the students' first and last names or any other unique labeling scheme. They should make two sets so each student will need two labels. Emphasize that they should not open the bags containing their set elements. Allow them class time to complete Part II of the activity.
OPTIONAL: Give the sets you made to the groups. The students will complete only Part II of the activity. You will again need to discuss how to write a set and make sure the students understand the activity. Have them switch sets with another group and complete Part II again with the new sets.

Grading Tips: Group or Combined Grade; 5 points for bringing three sets to class; 5 points for each correct union and intersection; 2 or 3 points for each set listed correctly (check listing against those of other team members)

Comments: If you decide to make the sets, be sure that groups of two or three students will have two sets each. In a class of 30 students, you will need to make at least 20 sets.
Connections: This activity can be used in any math class in which you are teaching set union and intersection.
Spin-offs: This activity has been used in a class for elementary school teacher preparation. It is a good example of how to use simple and inexpensive manipulatives in the classroom to teach a mathematical concept.

Activity 3-3: Working with Sets

Intermediate Algebra

Part I

Make two sets from the lists below, putting the objects in see-through, closable bags. You may pick items from any of the following three lists, but make sure all items in one set come from the same list. You need only put in five or six things, but use only one of each item because in sets you only list an item once no matter how many may be in the set.

Kitchen Items	Outside Items	Car Items
an article of silverware	a rock or stone	air freshener
an article of plastic tableware	grass clippings	a napkin or tissue
a napkin or paper towel	a twig or stick	something from the glove box
a measuring spoon or cup	a leaf	oil rag
a bottle opener	a flower	car key
a pen or pencil	paper trash	a pen or pencil
a tea bag	smashed aluminum can	any other small item
any other small item	any other small item	

Part II

Label your sets as the instructor directs so that each set has a unique label. Form groups and then complete the following steps. **DO NOT OPEN THE BAGS!**

1. Using roster or set-builder notation, each member of the group will list the elements of his or her two sets, associating the list with its label.

2. Each member will find the union of his or her pair of sets. Then the group members will check one another's work to make sure all unions are done correctly.

3. Now, working together, the group will find all intersections of their sets, making sure to use all combinations of pairs of sets for the group.

4. Make sure everyone gets his or her original sets back.

Activity 3-4: Building a Road

Intermediate Algebra

Slope (Grade)

NOTES TO THE INSTRUCTOR

Summary: Students determine grade and distance traveled, given the grade for a simple highway design.

Skills Required: Slope
Skills Used: Application of Slope and Grade

Grouping: Structured Groups, Ability Mixing or Matching
Materials Needed and Preparation: Copies of Activity 3-4, one for each group; pieces of cardboard, rulers, or other measuring devices; plan Structured Groups
Student Time: [**In Class**] 10 to 30 minutes; [**Out of Class**] group homework assignment

Teaching Tips: After some discussion on the slope of a line, have the students form their groups and complete the activity. Help them figure out how to model the grade (slope) with the cardboard. Assign the rest of the activity as homework, if needed.

Grading Tips: Group Grade; 10 points for questions 1, 3, 4, 6, and 7; 25 points for models from questions 2 and 5

Comments: Many students enjoy making the cardboard models of the slope. Be ready for some creative work and inaccuracies in the models.
Spin-offs: Display the best of the cardboard models with the student solutions in a prominent place.

Activity 3-4: Building a Road

Intermediate Algebra   Slope (Grade)

Your company has contracted to build a highway from one town to another and you are on the engineering design team. One portion of the road must go from a river valley to a ridge top. The surveyors tell you that there is a vertical rise of 150 feet from the valley to the top of the ridge. Your team decides to allow no more than a 6% grade on the road and that the uphill portion will be straight without any curves or switchbacks.

1. What horizontal distance would one travel going from the valley to the ridge with the 6% grade? Express your answer in feet and in miles (there are 5280 feet in a mile).

2. Use actual distances to make a three-dimensional, scale model of the 6% grade using cardboard or stiff paper and observe how objects roll down it.

One of the members of your team goes back to the surveyor's drawing and notices that the distance from the edge of the river to the base of the ridge is only 750 feet. Your team would like to minimize the amount of cutting and filling you will have to do, so you would like the middle of this uphill portion of the road to coincide with the ridgeline (see figure below).

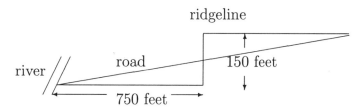

3. Is it possible, with a 6% grade, to have the middle of this uphill portion coincide with the ridgeline, given that you start your climb *after* crossing the river? (Show your work and explain your reason.)

4. What would the grade need to be to have the road fit the picture above, i.e., to have the middle of this uphill portion coincide with the ridgeline?

5. Make a three-dimensional model of this new grade with cardboard as you did in question 2. Compare the new model with your previous one and see whether objects roll down the grade more quickly or slowly.

6. What do you think the maximum safe grade should be for highways? Use your cardboard to make various models of different grades to determine this. Give an explanation for your group's choice.

Now, using the maximum safe grade you determined in question 6, you will need to design your road.

7. What horizontal distance would one travel going from the valley to the ridge with the maximum safe grade your group decided upon in question 6?

NOTES TO THE INSTRUCTOR

Summary: Through graphical exploration, students discover the properties of parallel and perpendicular lines.

Skills Required: Graphing, Finding Equations of Lines
Skills Used: Parallel and Perpendicular Properties

Grouping: Proximity Pairing (2 students per group)
Materials Needed and Preparation: Copies of Activity 3-5, one for each group; graph paper (students provide)
Student Time: [**In Class**] 50 minutes; [**Out of Class**] group homework assignment, if necessary

Teaching Tips: Have students decide which of them will be the Grapher first and which one will be the Answerer in each pair in the activity. Allow them enough time to complete the activity in class. Remind them to switch roles for Part II. This activity is designed to replace the lecture on parallel and perpendicular lines, but you may want to augment the activity with some class discussion.

Grading Tips: Group or Combined Grade; each question worth 4 points; 8 points for quality of graphs

Comments: Students enjoy having a class period to work one-on-one with another student. It is a good break from listening to lecture or trying to "get a word in edgewise" in a larger group. You may want to give some thought about some of the pairings, especially avoiding pairing a very forceful student with a student who is shy or lacks confidence.
Connections: Activity 3-11: *Complex Numbers;* Activity 3-14: *A Carnival of Conics—Parabolas;* Activity 3-15: *A Carnival of Conics—Circles and Ellipses;* and Activity 3-16: *A Carnival of Conics—Circles, Ellipses, and Hyperbolas* also provide a graphical understanding of the concepts used.

Activity 3-5:   Parallel and Perpendicular Explorations

Intermediate Algebra   Parallel and Perpendicular Lines

Part I

One person in your pair will do the graphing (**G**: Grapher) while the other one is answering the questions (**A**: Answerer). Be sure you discuss the answers and that you both know how to find them. Make sure to use a large enough coordinate plane to graph four or five lines on it. Also, make sure all points and lines are labeled on your graphs.

1. **G**: Graph the line through the points $(2, 3)$ and $(-1, 4)$. We will call this line S.   **A**: Find the equation for line S.

2. **G**: Plot points 3 units up (positive y direction) from each of the points given above on line S.   **A**: What are the new points?

3. **G**: Draw a line through the new points. We will call this line P.   **A**: Find the equation for line P.

Now, we will compare these lines. You will need to work together on these questions.

4. **G, A**: Looking only at your graph, state how these lines are the same and how they are different.

5. **G, A**: Make sure your equations are in slope-intercept form. State how the equations are the same and how they are different.

Let's look at another line to see if moving up (or down) is the only way to get this result.

6. **G**: On the same graph, plot points 5 units to the left (negative x direction) from each of the points given on line S.   **A**: What are the new points?

7. **G**: Draw a line through the new points. We will call this line L.   **A**: Find the equation for line L.

8. **G, A**: State how line L is the same as and different from lines P and S.

Now let's see what happens if we move a different number of units from our two original points.

9. **G**: Plot a point that is 1 unit left of $(2, 3)$ and a point that is $\frac{1}{2}$ unit right of $(-1, 4)$. Draw a line through the new points. We will call this line R.   **A**: Find the equation for line R.

10. **G, A**: State how line R compares to the other three.

Finally, it is your turn to use the information above to come up with another parallel line.

11. **G, A**: Find another line parallel to line S above. Graph it and state its equation.

Part II

At this point you should switch roles. Get a new sheet of paper and look at another way lines are related. Again, make sure that all points and lines are labeled on your graphs.

12. **G:** Graph the line through the points $(3,1)$ and $(-1,-2)$. We will call this line T. **A:** Find the equation for line T.

13. **G:** Plot one point 4 units down (negative y direction) and 3 units to the right (positive x direction) from $(3,1)$. **A:** What is this point?

14. **G:** Draw the line connecting $(3,1)$ and your new point. Call this line Q. **A:** Find the equation for line Q.

Now, we will compare these lines. You will need to work together on these questions.

15. **G, A:** Looking at the graph, state how the lines are related. State how the equations for the two lines compare.

16. **G, A:** Look at question 13; explain how the direction and number of units you moved from $(3,1)$ relates to the slope of your new equation.

Now, let's see if we can make another perpendicular line.

17. **G:** Plot the point $(-4,2)$ and draw a line from the point $(-1,-2)$ to this new point. Call this line M. **A:** Find the equation for line M.

18. **G, A:** Looking at the graph, state how line M relates to lines T and Q. State how their equations compare.

Now we will see if moving from the point $(3,1)$ in a different way will give us a perpendicular line.

19. **G:** Plot one point 4 units down (negative y direction) and 1 unit to the left (negative x direction) from $(3,1)$. **A:** What is this point?

20. **G:** Draw the line connecting $(3,1)$ and your new point. Call this line F. **A:** Find the equation for line F.

21. **G, A:** State how line F relates to lines T, Q, and M. State how their equations compare.

Finally, we will draw some conclusions about perpendicular lines.

22. **G, A:** Starting from any point on line Q, other than $(3,1)$ and $(-1,-2)$, how many units left or right (negative or positive x direction) and how many units up or down (positive or negative y direction) would you need to move to get to a point that would be on a line perpendicular to Q? Explain.

23. **G, A:** Find another line perpendicular to line T above. Graph it and state its equation.

NOTES TO THE INSTRUCTOR

Summary: Students will fill out tables and answer questions to help them determine the percent of juice concentrate in a container of juice mixture.

Skills Required: Solving Simple Linear Equations, Percents
Skills Used: Application Problems Involving Percent Solution

Grouping: Proximity Pairing or Count Off
Materials Needed and Preparation: Copies of Activity 3-6, one for each group; at least one small can (6 oz) of juice concentrate, a container with enough water to make the juice, a container in which to make the juice, and a tray to catch spills; cups for students to drink the juice afterwards; OPTIONAL: Have students bring the juice and empty container and the instructor supply water and cups.
Student Time: [**In Class**] 30 minutes; OPTIONAL: 5 minutes on Day 1, 30 minutes on Day 2; [**Out of Class**] none; OPTIONAL: group homework assignment to get supplies

Teaching Tips: This activity is designed to augment the lecture on solving percent-solution mixture problems. Mix the juice while the students are answering the appropriate questions and filling out the tables. Break the students into groups and circulate around the room while they are completing each portion of the activity. The activity is designed to walk the students through the solution, but they may need some help in setting up their equations from the table. OPTIONAL: Form students into groups during the class period before covering mixture problems. Give them a few minutes at the end of class to determine what type of juice to bring and who will bring the juice and container. Have each group bring a small (6 oz) can of juice concentrate of their choice and a container large enough to hold the reconstituted juice. Provide water, cups, and paper towels to wipe up spills for the students. Each group will perform the demonstration themselves as they fill out the questions on the activity.

Grading Tips: Individual or Group Grade; 5 points each for questions 1 and 3; 10 points for questions 2 and 4; 25 points for question 5 distributed with 5 points for the table, 10 points for the equation, and 10 points for the answer; 45 points for question 6 distributed with 5 points for the table, 15 points for the equations, and 25 points for the answer

Comments: Orange juice is a solution with which almost all students are familiar. This will make the percent solution problems less obscure than the typical alcohol/water or acid/water chemistry type problems. It is also designed to be a visual demonstration so students can see that the concentrate is separate from the water until mixing. This visualization will help them understand how there can be 1 can of juice concentrate in 4 cans of solution.

Activity 3-6: Orange Juice Demonstration

Intermediate Algebra   Application Problems Involving Percent Solution

Group Members: _____

_____

Answer the questions below during the demonstration.

Given one can of juice concentrate and three cans of water, mix these together.

1. How many total cans of solution do you have?_____

2. What percent of the solution is juice concentrate?
    Use the space below or your own paper to show your work.

3. Would you need to add water or juice concentrate to make a 20% solution?_____

4. How many additional cans? (The table below will help you organize the given information. You may want to use proportions.)

	Original juice mixture	New juice mixture
Amount of solution		
Amount of juice concentrate in solution		

Use your own paper to show your work.

5. In a container that will hold 9 cans of liquid, suppose you mix 4 cans of a 25% solution and all of the 20% solution from question 4. Determine the percent of juice concentrate in the resulting solution. (The table below will help you organize the given information.)

	Original juice mixture	20% juice mixture	New juice mixture
Amount of solution			
% of juice concentrate		20%	
Amount of juice concentrate in solution			

Use your own paper to show your work.

166

Activity 3-7: Coffee on the Run!

Intermediate Algebra   Applications of Systems of Equations and Inequalities

NOTES TO THE INSTRUCTOR

Summary: Students are given a production problem for which they will perform a cost analysis by solving systems of inequalities in two variables.

Skills Required: Methods for Solving Systems of Equations, Graphing Inequalities in Two Variables
Skills Used: Applications of Solving Systems of Equations and Inequalities

Grouping: Structured Groups, Ability Mixing or Matching
Materials Needed and Preparation: Copies of Activity 3-7, one for each student; copies of Activity D: *Roles for Groups,* one for each group, if needed; plan Structured Groups if not already established
Student Time: [**In Class**] 20 minutes; [**Out of Class**] group homework assignment

Teaching Tips: This activity is designed to take several days and require group work outside of class. Hand out the activity when you start the unit on solving systems of equations in two variables. Students will be able to complete questions 1, 2, and 3 before you have covered all of the material on solving systems of equations. Have the students read the entire activity through before beginning work. They will need to pay close attention to the information provided in the activity. The messenger in each group can come to you for conversion information, given below, to help them fill out the table. Make sure they have defined the correct variables (cups of each type of coffee produced) before they continue on the activity, or they may not be able to complete it. Have parts of the activity due each class period so that the students work on it over the entire time allotted rather than waiting until the night before the final report is due to start. OPTIONAL: This activity could be given out close to the beginning of the semester as a quarter or midterm project. The material needed to complete the first part of the activity is usually covered early in a course. Portions of the activity could then be due with the test on the corresponding material. The final activity would then be turned in at midterm or the end of a quarter.

Grading Tips: Group Grade; 3 points each for questions 1, 4, 6, 8, 10, and 13; 8 points each for 2, 3, 9, 11, and 12; 20 points each for 5 and 7; 8 points report quality

Comments: If students are having trouble keeping track of all the information, suggest that they make a list of all the known values. They will be able to use many of the options available on programmable calculators with this activity.
Spin-offs: Some students may be interested in putting this simple, linear programming problem into the appropriate computer software (for example, a program that uses the simplex method for solving such problems). The activity could also lead to discussions in linear algebra.

Conversion Information: There are 8 fluid ounces in one cup, 128 fluid ounces in one gallon, and 16 ounces in one pound.

Activity 3-7:   Coffee on the Run!

Intermediate Algebra   Applications of Systems of Equations and Inequalities

Your group has decided to sell coffee to commuters in the heavy traffic near your school. You decide to do a cost analysis to determine whether to make both caffeinated and decaffeinated and how many cups of each type to make. There are two types of coffee grounds, and each type of coffee requires water, so you will have a total of three raw materials with which to work. You decide to use bottled water to ensure purity.

1. There are two types of coffee you want to produce. Choose a variable to represent each type. Give a complete definition of what each variable represents and use cups of coffee as your units.

2. Using the following information, fill in the table below with the amount of the raw materials needed on a daily basis to make the coffee.

    a. You decide to sell 8-ounce cups of coffee, and one person in the group has volunteered a 10-cup coffee maker.

    b. The coffee maker you will be using requires 2 ounces of caffeinated or $1\frac{1}{2}$ ounces of decaffeinated coffee grounds for the whole pot.

    c. Your group has enough money to buy a maximum of two pounds of caffeinated coffee, one pound of decaffeinated coffee, and two 5-gallon containers of bottled water. (Convert these to ounces before filling in the table!)

Raw material	Amount needed for caf. per cup	Amount needed for decaf. per cup	Max. amount of raw material available
Bottled water (fluid ounces)			
Caf. coffee grounds (ounces by weight)			
Decaf. coffee grounds (ounces by weight)			

The values from part c, your maximum amounts of raw material available, will determine your inequalities for the cost analysis. The inequalities set limits on your variables, in this case, the amount of money you could invest in start-up supplies.

3. Write the correct inequalities for each raw material you will be using from the information in your table. Each row of the table will give you an inequality.

4. Now find inequalities to show that you will make zero or more cups of each type of coffee. These inequalities close the bounded region so your cost analysis will give you nonnegative numbers. (There are times when the best option is to make nothing.)

You should now have five inequalities.

168

5. Graph the inequalities you found in questions 3 and 4, using your two variables as the axes of the coordinate plane. Graph all the inequalities on the same coordinate plane. Shade the region that forms the intersection of all of your inequalities.

The graph of these inequalities should form a **bounded region;** in other words, the lines forming the boundary for each inequality will form the boundaries for an enclosed space. Where these lines cross will determine how much coffee you will make.

6. List the lines that form the boundaries of the graph you made in question 5. These lines should come directly from the inequalities you found in questions 3 and 4.

7. Use any method of solving systems of equations to find the points where these lines cross (the vertices of the bounded region). One of the points you find should be the point $(0,0)$. You will have a total of six points.

8. Two of the points you found in question 6 are outside the shaded region you graphed in question 5. Find which points they are and list them.

So far you have found *possible* quantities of coffee to produce. As any businessperson will explain, you don't stay in business unless you make a profit. So, your group will want to know which production choice will maximize profits. Suppose you decide to sell both types of coffee at $0.50 per cup. Your raw materials have the following costs:

Caffeinated coffee: $0.14 per ounce.

Decaffeinated coffee: $0.28 per ounce.

5-gallon bottle of water (excluding cost of bottle): $0.40 per gallon.

9. Find the cost per cup of coffee for each type of coffee. (*Carry as many digits in your answer as you can.*) You may wish to make a table for these values.

10. What will be your net profit (selling price less cost) per cup of coffee for each type? (*Carry as many digits in your answer as you can.*)

11. Set up an algebraic expression showing your total net profit for selling both caffeinated and decaffeinated coffee.

12. Using the points that are the vertices of the bounded region, plug in the amounts of each type of coffee you might produce. You will get four possible profit scenarios. (*Round your answers to the nearest cent.*) Which of the four possible scenarios gives you the maximum profit? Which of the scenarios is second best?

You have now determined the amount of each type of coffee to produce based on cost and availability of raw materials. If you were to try to do this in reality, you might decide to go with slightly lower profits to provide a variety.

13. Discuss which profit scenario *your* group would choose: whether to make both types or just one and how much, based on the cost analysis. State the reasons for your choice.

NOTES TO THE INSTRUCTOR

Summary: Students are asked to solve quadratic equations given a real-life problem of building a sunroom/studio onto a house. They use area and building code considerations to determine dimensions and feasibility of the construction.

Skills Required: Solving Quadratic Equations by Factoring
Skills Used: Applications of Quadratic Equations

Grouping: Structured Groups, Ability Mixing or Matching
Materials Needed and Preparation: Copies of Activity 3-8, one for each group; Activity D: *Roles for Groups,* one for each group, if needed; plan Structured Groups
Student Time: [**In Class**] 15 to 30 minutes; [**Out of Class**] group homework assignment

Teaching Tips: This activity will require the groups to meet outside of class. You will want to give them at least 15 minutes in class to assign roles and tasks and to set up a meeting time outside of class. They may also want to finish their drawing (question 1) and set up the expressions for question 2 before they leave class. Remind them to call the correct agency to answer question 4 about the building code. Have students turn in group or individual reports of their solutions. OPTIONAL: You may decide to give them building code information rather than have them call. The building codes may be different in different cities or counties. Below is information they may use:

> **Building Code Information** (use if students do not call for the information):
>
> 7 feet required between any structure and a side of the lot
>
> 25 feet required between any structure and the back of the lot

Grading Tips: Group Grade; 10 points for questions 1 and 2; 15 points for questions 4, 5, and 6; 20 points for question 3; 15 points report quality

Comments: This activity provides a good exercise for the students to pick the important information out of a story or "real-life" situation. If you structure the groups using Ability Mixing, the more able students can help their teammates set up and solve the problem.

Activity 3-8:   Building a Sunroom

Intermediate Algebra   Solving Quadratic Equations by Factoring

You have found a summer job in Santa Fe working on a construction crew. One of the projects on which your crew will work is building a rectangular sunroom/studio on the south side of an adobe house. The studio is attached so that the north side of the sunroom will be a portion of the current south side of the house, and the west wall of the new room is to be made flush with the current west wall of the house. It is to be a rammed-earth room in order to take advantage of solar heating, so the walls of the sunroom will be 2 feet thick.

1. In your groups, assign each member a role, either Moderator, Quality Manager, Recorder, or Messenger. Then sketch a drawing of the house and approximately where the sunroom should go. Leave room in your drawing for more information.

Because of the cost of building materials and labor, based on what the contractor suggests, the people who own the house decide to limit the total area of the sunroom/studio. They also decide to make the inside south wall twice as long as the inside west wall.

2. Add the new information to your drawing and define your variable(s).

3. What are the internal dimensions if the *external* area is 392 square feet? Find an algebraic equation for the external area using the variable(s) you defined in question 2.

Call the appropriate agency and determine the distance between a structure and the property lines the building code requires in your city. This house currently (before the construction) sits 40 feet from the west property line (the back of the lot) and 22 feet from the south property line (a side of the lot).

4. List the building code requirements you found.

5. What are the external dimensions of the room?

6. Will the sunroom stay within code if you build it as it is designed now?

Activity 3-9:   Making a Bid

NOTES TO THE INSTRUCTOR

Summary: Students use rational equations to solve generalized work problems in the setting of bidding on landscaping jobs.

Skills Required: Setting Up Work Problems, Proportional Reasoning
Skills Used: Application of Work Problems, Generalized Work Problems

Grouping: Structured Groups, Ability Mixing or Matching
Materials Needed and Preparation: Copies of Activity 3-9, one for each group
Student Time: [**In Class**] 20 to 30 minutes, on each of 2 days; [**Out of Class**] group homework assignment

Teaching Tips: After explaining how to solve simple work problems, assign the students this activity to solve in their groups, rather than solving the application problems in the textbook. They will need at least two class periods to work on the activity; Part I started (and possibly completed) on Day 1, and Parts II and III on Day 2. You may want to have Part I due the day on which they work on Parts II and III. Remind the students to make the calls needed to answer questions A and B before starting to work on Part II. The problems in this activity are presented in a different way from traditional work problems. The activity is designed to lead students to a better understanding of the general form of work problems. Work the activity before assigning it to the students so you will be able to help them take the steps needed to make this generalization.

Grading Tips: Group Grade; 10 points each for questions 1, 2, 3, and 4 (total 40 points); 15 points for calling a nursery or landscaping company; 15 points each for questions 5, 6, and 7

Comments: This activity does not have all of the steps for solving it spelled out for the students. They will need to brainstorm about how to tackle parts of it.

Activity 3-9:   Making a Bid

Intermediate Algebra   Solving Work Problems

Part I
You and your group are trying to start a landscaping business and you want to be able to bid on jobs. You decide you need to know how long it will take to do any particular job and how much it will cost. One of the things you will be doing is planting trees. After some consultation, you find out that, in good soil, one member of your group (say GM1) can plant 5 trees in 2 hours.

1. At this rate, how long would it take GM1 to plant 30 trees?

You also find that another member of the group (GM2) has planted trees with GM1 and working together they planted 12 trees in 3 hours.

2. Using this information and GM1's rate from above, what would GM2's rate be?

3. How long will it take them to plant 30 trees if a third group member (GM3) helps and he or she can plant 4 trees in 2 hours?

Part II
Another job you will do is put down sod. Call a landscaping company or nursery to find out how long it takes its crew to put down sod. Do this by asking the following questions:

> A. How long would it take the crew to put down 4500 square feet of sod?
> B. How many people do you typically have on a crew?

4. Find the typical working rate per person for putting down the sod $\left(\frac{A}{B}\right)$.

After some discussion, you decide that one of your group members (GM4) will be in charge of the bookkeeping and will only help occasionally on the landscaping work. The first job on which you wish to bid will require putting down 6000 square feet of sod. GM4 plans to help the rest of you for only half of the sod job. Assume that all four of you (GM1, 2, 3, and 4) can work at the rate you found in question 4.

5. Find an equation for how long it will take all four of you working together to put down half of the sod. Define your variable(s) and solve the equation.

6. Find an equation for how long it will take three of you to put down the rest of the sod. Define your variable (Is it different from the one in question 5?) and solve the equation.

Part III
You want to figure out how much it will cost to plant trees and put down sod. Assume that you will all receive $6.00 per hour for both planting trees and putting down sod.

7. Make your bid on the labor part of the job to plant 30 trees and put down 6000 square feet of sod as discussed above (question 3 and questions 5 and 6).

NOTES TO THE INSTRUCTOR

Summary: Students are given proportion problems in which they finalize some of the design considerations for a water works for a learning park.

Skills Required: Knowledge of Ratio and Proportion
Skills Used: Applications of Ratio and Proportion

Grouping: Structured Groups, Ability Mixing or Matching
Materials Needed and Preparation: Copies of Activity 3-10 and Activity D: *Roles for Groups,* one for each group
Student Time: [**In Class**] 15 to 20 minutes; [**Out of Class**] none

Teaching Tips: This activity is designed to be a way to bring the use of proportion equations into a setting the students may encounter in a service or parent–teacher organization. Have the students assign roles in their groups (see Activity D) before they begin work on the activity. Allow the students class time to share their answers with the entire class and to discuss any other design considerations.

Grading Tips: Group Grade; 22 points each for questions 1 through 4; 12 points for question 5

Comments: Students may wish to take more class time to finalize the design of their water works.
Spin-offs: Have students brainstorm with engineering students to finalize the design and then construct a working model of the water works as described in this activity. Have students visit local science or children's museums.

You are involved with an organization in your town that has decided to design and build a hands-on learning park for children. Included in the park will be a water works to demonstrate how water can do various things for people as well as other scientific exhibits, all of which can be changed and manipulated by the children who come to the park. You are working with the group that is setting up the water works. Below are some design considerations you and your group need to finalize.

The water works will begin with a holding tank that has a hole at the bottom to let the water out and a pipe at the top through which the recycled water will go back into the tank. You plan to have a model water wheel and a model hydroelectric generator run by the water coming down a channel.

1. The holding tank is a metal, 50-gallon drum donated by an area business. The inside of the tank is coated so it will not rust, and one member of your group has cut a hole in the bottom. You want to find out at what rate the water will come out of the tank through the hole. You assume a constant rate of flow and fill the tank half full of water (actually, the rate is dependent upon how full the tank is). Another member of the group times the water flow and finds that it takes 6.25 minutes to empty the tank. Find the rate of water flow in gallons per minute. Using a proportion equation, find how long it would have taken to empty the tank if it had been full.

2. A member of your group has obtained a small, used water pump from a local landscaping company. You wish to find out at what speed the pump will recirculate the water. You find that it takes 2.25 minutes to pump 10 gallons into the holding tank. At what rate, in gallons per minute, will the pump recirculate the water? How long would it take to fill the holding tank?

You have found that the angle of the channel determines how fast the water flows through it. You want the children to be able to move the channel up or down to allow the water to go faster or slower so they will see how that affects the water wheel and the generator. You decide to have the flow rate be adjustable to a half a gallon per minute faster or slower than the rate you found in question 1.

3. Putting the channel in the most upward position (flow rate will be slower), what do you want the rate to be? You time the water flow for 5 minutes through the channel. Using a proportion equation, find how many gallons should have passed down the channel if it is flowing at the rate you want.

4. Now, put the channel in the most downward position (flow rate will be faster), and repeat question 3 for this rate.

5. After completing the problems above, draw a diagram of how your group would put the water works together, including the water wheel and generator.

NOTES TO THE INSTRUCTOR

Summary: Students are given a graphical representation of complex numbers that is related to the algebraic representation of the numbers, the operations of addition and subtraction of complex numbers, and the complex conjugate.

Skills Required: Graphing, Complex Numbers
Skills Used: Complex Numbers, Addition, Subtraction, Conjugate

Grouping: Proximity Pairing
Materials Needed and Preparation: Copies of Activity 3-11, one for each group; graph paper (students provide); copies of Activity C: *To the Student,* one for each group, if needed
Student Time: [**In Class**] 20 to 30 minutes; [**Out of Class**] group homework if needed

Teaching Tips: This activity is designed to augment the lecture on complex numbers and to introduce the concept of complex numbers in a graphical way. Assign this as an in-class group activity on the day you plan to teach complex numbers. Plan on allowing the students class time to finish Part I and assign Part II as homework. Have copies of Activity C available in case the groups need help working together.

Grading Tips: Group Grade; 15 points for each of the six questions and 10 points for solution quality

Comments: Students, like mathematicians in the past, find the association of complex numbers to something with which they are already familiar (the xy-coordinate plane) gratifying and demystifying. Giving them a look at the uses of complex numbers helps to motivate them and to encourage them to take the subject seriously.
Connections: Activity 3-5: *Parallel and Perpendicular Explorations;* Activity 3-14: *A Carnival of Conics—Parabolas;* Activity 3-15: *A Carnival of Conics—Circles and Ellipses;* and Activity 3-16: *A Carnival of Conics—Circles, Ellipses, and Hyperbolas* also provide a graphical understanding of the concepts used.

The complex number system is fairly new, as number systems go, having been developed in the last 200 years. The complex numbers were not at first accepted by all mathematicians. When a geometric representation was found for them, however, people could understand their usefulness and the complex numbers came into common mathematical usage. The geometric representation of the complex numbers that was first developed associated each complex number with a point in a plane. We will look at that representation in this activity.

Part I
First we will do a quick review.

1. Plot these points in the xy-plane, labeling axes, indicating a scale, etc.

    a. $(2, 3)$ b. $(-3, 5)$

    c. $(-2, -1)$ d. $(6, -2)$

    e. $(3, 0)$ f. $(0, -5)$

Complex numbers have two parts, a **real part** and an **imaginary part**. They are usually written like this: $x + yi$, where x is the real part and yi is the imaginary part. Now, we associate the y-coordinate in each ordered pair in the xy-plane with the imaginary part and the x-coordinate with the real part of a complex number. We then have a relationship between points in the plane and the set of complex numbers.

Example: $(1, 3)$ represents $1 + 3i$

2. Associate the ordered pairs in question 1 with complex numbers. Notice what happens to points that lie on an axis.

The operations performed on the two representations are similar. When adding two complex numbers, we add the real parts together and the imaginary parts together:

$$
\begin{array}{r}
3 \;+\; 5i \\
+\; 4 \;+\; 6i \\
\hline
7 \;+\; 11i
\end{array}
$$

Subtraction is similar; simply add the additive inverse:

$$
\begin{array}{r}
3 \;+\; 5i \\
-\; (4 \;+\; 6i) \\
\hline
\end{array}
\qquad \text{becomes} \qquad
\begin{array}{r}
3 \;+\; 5i \\
+\; (-4 \;-\; 6i) \\
\hline
-1 \;-\; i
\end{array}
$$

We can perform similar operations with ordered pairs in the xy-plane. When plotting the point $(3, 5)$, we start at $(0, 0)$ and move 3 units in the positive x direction and 5 units in the positive y direction to arrive at the point in the plane. To "add" the point $(4, 6)$, we would start at the point $(3, 5)$ and move 4 units in the positive x direction and 6 units in the positive y direction to arrive at the resulting point of $(7, 11)$. To subtract, reverse the directions for both the x- and y-coordinates. So to subtract $(4, 6)$, move 4 units in the *negative x* direction and 6 units in the *negative y* direction.

3. Compute the sums (and differences) below and then show the graphical representation of each one. Remember that subtraction is just the same as adding an additive inverse.

    a. $(7 + 9i) + (3 - 2i)$ b. $(2 - 5i) + (-4 + 3i)$

    c. $(-2 + 6i) + (3 - 7i)$ d. $i + (6 - i)$

    e. $(-1 + 3i) - (2 - 2i)$ f. $(-2 - 6i) - (-2 + 3i)$

Part II

The other association of interest is that of the **complex conjugate.** Algebraically, to find the conjugate of a complex number, we change the sign of the imaginary part. What does this mean graphically?

4. Find the conjugates for the complex numbers below.

	Complex Number	Complex Conjugate
a	$9i$	
b	$3 - 4i$	
c	$-1 + 3i$	
d	10	

5. Associate the appropriate ordered pairs with each complex number and its conjugate and graph them. Plot one pair of points (complex number and its conjugate) on each graph.

6. Draw a line from the origin to each point. Describe how these lines are related.

The lines from the origin to the complex number and its conjugate are **reflected** across the x-axis. In higher-level mathematics courses and in science and engineering this is an important relationship. You will also see complex numbers being associated with **vectors.**

Activity 3-12:   Not That Sunroom Again!

NOTES TO THE INSTRUCTOR

Summary: Students are given a problem related to the world of home remodeling in which they will use the quadratic formula to solve for the dimensions of a sunroom addition.

Skills Required: Area of a Circle
Skills Used: Application of Quadratic Equations

Grouping: Structured Groups, Ability Mixing or Matching
Materials Needed and Preparation: Copies of Activity 3-12, one for each group; copies of Activity D: *Roles for Groups,* if needed
Student Time: [**In Class**] 20 minutes; [**Out of Class**] group homework assignment

Teaching Tips: Hand out this activity when you start the unit on solving quadratic equations using the quadratic formula. Students will be able to complete questions 1 and 2 before you have covered the quadratic formula. You may want them to read the activity and conjecture a way to solve the problem before covering the material needed. Mention to students that rounding intermediate answers can cause their final answer to be off a little bit. Announce whether the students will hand in individual or group reports.

Grading Tips: Individual or Group Grade; 10 points for question 2, 25 points for question 3, 50 points for question 4, 15 points for report quality

Comments: Discuss in class how mathematics can model reality and how much time it can save. Have students do a scale drawing and mark in and count the tiles.
Connections: Activity 3-8: *Building a Sunroom*

Activity 3-12:   Not That Sunroom Again!

Intermediate Algebra   Solving Quadratic Equations by the Quadratic Formula

It is almost the end of the summer in Santa Fe. You will be able to work on one more construction job before you go back to school. On this next job, the owners of the house had heard about a sunroom your crew built earlier in the summer for some friends of theirs and they want a sunroom addition for their house as well. The sunroom is to be a semi-circle and the walls are to be rammed earth. There will also be a 4-foot-wide, semi-circular patio around the sunroom. The homeowners have bought 400, 12-inch square, Saltillo tile while in Mexico for their summer vacation. They want the dimensions of the sunroom to be such that they can use the tile for the floor of the sunroom *and* for the patio. Below is a sketch the homeowners have given your boss and which he passes on to your crew to use in determining the dimensions of the new sunroom.

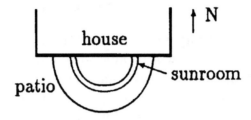

1. In your groups, assign each member a role, either Moderator, Quality Manager, Recorder, or Messenger.

Refer to the drawing above in your calculations. The rammed earth walls will be 2 feet thick. Use 3.14 as an approximation for π unless you have a calculator with a π key, in which case carry as many digits as your calculator will hold.

2. Add the above information to the drawing and define your variable(s).

3. The total area for both the inside of the sunroom and the patio will need to be less than or equal to the area covered by the tiles. Determine an equation or inequality to model this.

4. Solve your equation from question 3 using the quadratic formula and determine the radius of the semicircular sunroom. Round your answer to two decimal places. How long would the south wall of the house have to be in order for this design to work?

5. Make sure you show all of your work and state any assumptions you have made on the paper(s) you hand in. Your instructor will determine if you will hand in individual papers or a group paper.

NOTES TO THE INSTRUCTOR

Summary: Students are asked to find the solution set for rational and quadratic inequalities.

Skills Required: Solving Simple Linear Inequalities, Graphing Intervals on a Number Line
Skills Used: Finding and Graphing the Solution Set for Quadratic and Rational Inequalities

Grouping: Structured Groups, Ability Mixing or Matching
Materials Needed and Preparation: Copies of Activity 3-13, one for each group; fill in the original problem you wish them to work
Student Time: [**In Class**] 20 minutes; [**Out of Class**] none

Teaching Tips: This activity is designed to reinforce lectures on finding the solution set for quadratic and rational inequalities. You will need to fill in the problems on the activity sheet that you wish students to work, or refer them to problems in the book. Hand out the activity after covering the main points in lecture. This activity should be done in class so students may ask for help. Do an example with them so they can ask questions about what to do. An example appears on page 2 of the Notes to the Instructor.

Grading Tips: Group Grade; 5 points for factoring correctly if needed (question 1), 5 points for filling in factors and names correctly on the record sheet, 30 points for solving inequalities correctly (finding "break points" and intervals), 25 points for graphing intervals correctly on the number line, 25 points for the correct solution for the original problem, 10 points for including endpoints correctly; see example in answer key

Comments: This method of solving rational and quadratic inequalities may be unfamiliar to you. Finding the solution this way reinforces the rules of multiplying and dividing positive and negative numbers and helps the students see graphically why the solution is what it is.

NOTES TO THE INSTRUCTOR

Example:

Original Problem: $\dfrac{x^2 + 3x - 10}{x^2 - x - 56} \leq 0$

1. Factors: $(x+5)$, $(x-2)$, $(x-8)$, $(x+7)$

Name: _Olivia Sanchez_ **Factor:** _x + 5_ **Break Point:** _−5_

Inequalities: $x > -5$, positive $\qquad$ $x < -5$, negative

Name: _Oliver Suzuki_ **Factor:** _x − 2_ **Break Point:** _2_

Inequalities: $x > 2$, positive $\quad$ $x < 2$, negative

Name: _Orville Sondheim_ **Factor:** _x − 8_ **Break Point:** _8_

Inequalities: $x > 8$, positive $\quad$ $x < 8$, negative

Name: _Olive Sanders_ **Factor:** _x + 7_ **Break Point:** _−7_

Inequalities: $x > -7$, positive $\quad$ $x < -7$, negative

Graph:

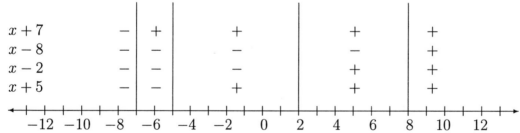

Solution: $\{x \mid -7 < x \leq -5 \text{ or } 2 \leq x < 8\}$ constitutes the set of x for which $\dfrac{x^2 + 3x - 10}{x^2 - x - 56} \leq 0$.

Activity 3-13: Getting Rational About Inequalities

Fill in the blanks below and on the record sheet while answering these questions.

Original Problem:

1. Together, list all of the factors from the problem above. Factor any portion of the algebraic expression that needs to be factored.

2. Each team member or pair of team members now picks one of the factors. Make sure all of the factors are picked and that each team member or pair has only one. List the factors on the record sheet and put your name(s) in the blank next to the factor you picked.

3. Each of you will now determine for which value your factor is zero. This value will determine your "break point" for the solution set. Fill in the break point for your factor on the record sheet.

4. Set up and solve inequalities to determine for which values the factor is positive and negative. Look at your original problem to determine whether the inequalities are strict inequalities or whether you should allow for equality also. You may also use test points to determine which side of the break point gives a positive or negative value.

5. Graph the solution set for each factor on one number line by drawing a vertical line at the break point for each and then indicating with $+$ and $-$ signs where the factor is positive and negative.

6. Determine in which interval(s) you would have a positive or negative product or quotient (see where the intervals you graphed in question 5 overlap). Remember that an even number of negative factors will give a positive answer and an odd number of negative factors will give a negative answer. Look back at your original problem. Did you want the quadratic or rational expression to be positive or negative? (Was it a $>$, $\geq$, $\leq$, or $<$ inequality?) Which endpoints do you need to include, if any? Write the solution to the original problem in set notation.

RECORD SHEET

Name:_____   Factor:_____   Break Point:_____
Inequalities:

Name:_____   Factor:_____   Break Point:_____
Inequalities:

Name:_____   Factor:_____   Break Point:_____
Inequalities:

Name:_____   Factor:_____   Break Point:_____
Inequalities:

Name:_____   Factor:_____   Break Point:_____
Inequalities:

Graph:

$$-12 \quad -10 \quad -8 \quad -6 \quad -4 \quad -2 \quad 0 \quad 2 \quad 4 \quad 6 \quad 8 \quad 10 \quad 12$$

Solution:

184

NOTES TO THE INSTRUCTOR

Summary: Students explore various aspects of the graphs and equations for parabolas using graphing calculators or computer software.

Skills Required: Use of Graphing Calculator or Computer Software
Skills Used: Comparison of Conic Sections

Grouping: Structured Groups, Ability Mixing, or Matching (2 students per group)
Materials Needed and Preparation: Copies of Activity 3-14, one for each group; graphing calculator for each group; plan Structured Groups
Student Time: [**In Class**] 30 to 45 minutes; [**Out of Class**] group homework assignment

Teaching Tips: This is the first of three activities designed to augment lectures on conic sections. Make sure each group has assigned roles and has access to a graphing calculator or a computer with the appropriate software loaded. You will need to spend some time the first day teaching the students the use of the graphing calculators or the computer software if you have not been using these already in your class. This activity is somewhat of a review of parabolas and can be used to teach the use of the calculators. It is to be completed in one day of class time. If groups need more time, you may (1) assign Part II of the activity as a group homework assignment, (2) complete Part II on an overhead, with the class giving input and predictions, or (3) assign each group a different pair of equations from Part II to complete and then report back to the class. Spend the last 10 or 15 minutes of class providing closure to the activity completed, having groups share what they have learned from the activity, and hitting the high points to consolidate learning. OPTIONAL: You may instead split this activity, in combination with Activity 3-15 and Activity 3-16, over two or three lab periods.

Grading Tips: Individual or Combined Participation Grade. If Part II is given as homework, this could be graded with points split appropriately.

Connections: Activity 3-15: *A Carnival of Conics—Circles and Ellipses* and Activity 3-16: *A Carnival of Conics—Circles, Ellipses, and Hyperbolas;* Activity 3-5: *Parallel and Perpendicular Explorations* and Activity 3-11: *Complex Numbers* also provide a graphical understanding of the concepts used.

Activity 3-14:   A Carnival of Conics—Parabolas

USE WITH GRAPHING CALCULATORS. Assign one person in your group the role of Recorder to write your answers on a separate paper. Assign the other person the role of Grapher to use the calculator or software to do the graphs. Switch roles for each problem so that each of you gets a chance to use the calculator or software. Calculator instructions are in *italic* type.

Part I

Before beginning the graphs, make sure your grid or coordinate axes are centered at $(0,0)$ and numbered *at least* to -10 and 10 in both the horizontal and vertical directions. (These should be the *standard* settings.)

1. *Graph* the quadratic equation, $y = 2x^2 - 20x + 53$. Put the cursor on the graph and *zoom* in and out to explore the shape of the graph. Try to describe the shape in words.

2. Return to the original graph and *trace* along the **parabola** until you are at the lowest point (watch your y-value changing until it is smallest). *Zoom* in a couple of times to get accurate x- and y-values for the lowest point. List the point you found. This point is called the **vertex.**

3. Return to the original graph. Then *zoom* out and *draw* a vertical line through the vertex. Describe how the parabola relates to the line. Determine the equation for the line. This line is called the **line of symmetry.**

Completing the square converts the equation above to **standard form** (or **vertex form**). The quadratic in standard form for this parabola is: $y = 2(x - 5)^2 + 3$.

4. Given the vertex you found in question 2, determine how you would find the x- and y- values for that point using the standard or vertex form of the equation. Also, determine how you would find the equation for the line of symmetry (question 3) by using the standard or vertex form of the equation.

For the following questions, keep only the standard form equation for the parabola above on your calculator.

5. On the same coordinate axes, graph each of the following parabolas. (You will have three parabolas graphed simultaneously.) Keep track of the order in which they are graphed so you will know which is the original. Describe how the new graphs are different from the original. Switch roles for each one so you will both have a chance to graph a parabola.

   a. $y = \frac{1}{2}(x - 5)^2 + 3$
   b. $y = -2(x - 5)^2 + 3$

6. Now clear your graph and, again, keep only the standard (or vertex) form of the equation for the parabola above. Graph the parabola, $x = 2(y - 5)^2 + 3$, and describe how it relates to your original parabola. (You will need to solve the equation for y, graph it in two parts, and *square* your graph.) What is the vertex and line of symmetry for the new parabola? How do they compare to the vertex and line of symmetry for the original parabola?

Part II

a. $y = 3x^2 + 12x + 8$

b. $y = -\frac{1}{3}x^2 - \frac{4}{3}x - \frac{16}{3}$

c. $x = 3y^2 + 12y + 8$

d. $x = -3y^2 - 12y - 16$

7. Convert the four equations above into standard (or vertex) form. State the vertex and line of symmetry for each.

8. Graph a and b on the same coordinate axes (have both equations turned on). Switch roles for each graph so each of you has a chance to graph one of the parabolas. Describe how a and b relate to each other.

9. Now, switch roles and graph equation c together with a only. (Remember to solve for y, graph the two parts, and square your graph.) Describe how a and c relate to each other.

10. Again, switch roles and graph equation d together with c only. (Solve for y, graph the two parts, and *square* your graph.) Describe how c and d relate to each other.

11. Describe how c and d relate to a.

NOTES TO THE INSTRUCTOR

Summary: Students explore various aspects of the graphs and equations for circles and ellipses and compare them using graphing calculators or computer software.

Skills Required: Use of Graphing Calculator or Computer Software
Skills Used: Comparison of Conic Sections

Grouping: Structured Groups, Ability Mixing, or Matching (2 students per group)
Materials Needed and Preparation: Copies of Activity 3-15, one for each group; graphing calculator for each group; plan Structured Groups
Student Time: [**In Class**] 30 to 45 minutes; [**Out of Class**] group homework assignment

Teaching Tips: This is the second of three activities designed to augment lectures on conic sections. Make sure each group has assigned roles and has access to a graphing calculator or a computer with the appropriate software loaded. You will need to spend some time the first day teaching the students the use of the graphing calculators or the computer software if you have not been using these already in your class. If groups need more time, you may (1) assign Part II of the activity as a group homework assignment, (2) complete Part II on an overhead, with the class giving input and predictions, or (3) assign each group a different pair of equations from Part II to complete and then report back to the class. Spend the last 10 or 15 minutes of class providing closure to the activity completed, having groups share what they have learned from the activity and hitting the high points to consolidate learning. OPTIONAL: You may instead split this activity, in combination with Activity 3-14 and Activity 3-16, over two or three lab periods.

Grading Tips: Participation Grade; if Part II is given as homework, this could be graded with points split appropriately

Connections: Activity 3-14: *A Carnival of Conics—Parabolas* and Activity 3-16: *A Carnival of Conics—Circles, Ellipses, and Hyperbolas*; Activity 3-5: *Parallel and Perpendicular Explorations* and Activity 3-11: *Complex Numbers* also provide a graphical understanding of the concepts used.

Activity 3-15: A Carnival of Conics—Circles and Ellipses

USE WITH GRAPHING CALCULATORS. Assign one person in your group the role of Recorder to write your answers on a separate paper. Assign the other person the role of Grapher to use the calculator or software to do the graphs. Switch roles for each problem so that each of you gets a chance to use the calculator or software. The computer instructions are in *italic* type.

Part I

Completing the square converts the equation, $x^2 + y^2 - 4x - 10y = -20$, to **standard form**. The standard form equation for this **circle** is: $(x - 2)^2 + (y - 5)^2 = 9$

1. *Graph* the standard form equation above. You will have to solve for y and enter the two equations you find. *Box* and *zoom* in the shape and *square* up your graph, to get the most accurate picture. Describe its shape.

2. *Draw* horizontal and vertical lines across the figure so they cross approximately in the **center** of the figure. Determine where the center point is. Determine the horizontal and vertical lengths across the graph by finding the endpoints of the lines you drew (where they intersect with the circle) and using the distance formula. Are these lengths the same?

The length of each of the line segments that you drew above is called the **diameter.**

3. Using the center point you found in question 2, determine how you would find this point using the standard form of the equation. The **radius** of the circle is half the diameter. Determine how to find the radius using the standard form of the equation.

Save your circle equations to graph again later, but *turn off* the graphs.

4. Now, graph the equation, $9(x - 2)^2 + 4(y - 5)^2 = 36$. You will again have to solve for y and graph the two equations you find. (*If you returned your graph to* **standard**, *you will need to box, zoom in, and square up your graph to get the most accurate picture.*) What is the shape of this graph?

5. Complete question 2 for your new graph.

6. Now, determine how you would find the center using the equation, $9(x-2)^2 + 4(y-5)^2 = 36$.

The standard form of the equation of an **ellipse** with center (h, k) is $\dfrac{(x-h)^2}{a^2} + \dfrac{(y-k)^2}{b^2} = 1$, $a > 0$, $b > 0$. To write the equation in question 4 in standard form, we must:

  a. Divide both sides by 36: $\dfrac{9(x-2)^2 + 4(y-5)^2}{36} = \dfrac{36}{36}$

  b. Simplify: $\dfrac{(x-2)^2}{4} + \dfrac{(y-5)^2}{9} = 1$

The equation is now in standard form: $\dfrac{(x-2)^2}{4} + \dfrac{(y-5)^2}{9} = 1$.

7. Describe how the numbers under the squared binomials compare to the lengths you found in question 5. These can be used to determine the lengths of the **major** (longer one) and **minor** (shorter one) **axes.**

Let's compare this ellipse to the circle we graphed previously.

8. Describe how the equation $9(x-2)^2 + 4(y-5)^2 = 36$ is similar to or different from the one for the circle: $(x-2)^2 + (y-5)^2 = 9$. Turn on all graphs (or graph both the circle and ellipse on the same axes) and compare them. Compare centers and lengths across the graphs. Describe what you discover.

9. How does the equation, $\dfrac{(x-2)^2}{9} + \dfrac{(y-5)^2}{9} = 1$, compare to the equation for the circle, $(x-2)^2 + (y-5)^2 = 9$? To the equation for the ellipse, $\dfrac{(x-2)^2}{4} + \dfrac{(y-5)^2}{9} = 1$? Conjecture about the relationship between circles and ellipses.

Part II

10. Return to standard axes. Write the following pairs of equations in standard form. State whether each is an ellipse or a circle, and determine the center of each, the radius of each circle, and the lengths of the major and minor axes of each ellipse. Graph each pair on the same coordinate axes. *Remember to box and square up your graphs!* Describe how each pair of equations compares. Switch roles in your groups for each pair of equations below.

    a. $9x^2 + y^2 = 9$ $x^2 + y^2 = 9$

    b. $25x^2 + y^2 = 25$ $x^2 + 25y^2 = 25$

    c. $4x^2 + \dfrac{1}{9}y^2 = \dfrac{4}{9}$ $4x^2 + y^2 = 4$

    d. $9x^2 - 18x + y^2 + 4y = -4$ $x^2 - 2x + y^2 + 4y = 4$

NOTES TO THE INSTRUCTOR

Summary: Students explore various aspects of the graphs and equations for circles, ellipses, and hyperbolas and compare them using graphing calculators or computer software.

Skills Required: Use of Graphing Calculator or Computer Software
Skills Used: Comparison of Conic Sections

Grouping: Structured Groups, Ability Mixing, or Matching (2 students per group)
Materials Needed and Preparation: Copies of Activity 3-16, one for each group; graphing calculator for each group; plan Structured Groups
Student Time: [**In Class**] 30 to 45 minutes; OPTIONAL: lab period; [**Out of Class**] group homework assignment

Teaching Tips: This is the third of three activities designed to augment lectures on conic sections. Make sure each group has assigned roles and has access to a graphing calculator or a computer with the appropriate software loaded. You will need to spend some time the first day teaching the students the use of the graphing calculators or the computer software if you have not been using these already in your class. If groups need more time, you may (1) assign Part II of the activity as a group homework assignment, (2) complete Part II on an overhead, with the class giving input and predictions, or (3) assign each group a different pair of equations from Part II to complete and then report back to the class. Spend the last 10 or 15 minutes of class providing closure to the activity completed, having groups share what they have learned from the activity and hitting the high points to consolidate learning. OPTIONAL: You may instead split this activity, in combination with Activity 3-14 and Activity 3-15, over two or three lab periods. You may also wish to add to this activity the task of graphing a hyperbola and its asymptotes on the same coordinate axes.

Grading Tips: Participation Grade; if Part II is given as homework, this could be graded with points split appropriately

Connections: Activity 3-14: *A Carnival of Conics—Parabolas* and Activity 3-15: *A Carnival of Conics—Circles and Ellipses;* Activity 3-5: *Parallel and Perpendicular Explorations* and Activity 3-11: *Complex Numbers* also provide a graphical understanding of the concepts used.

USE WITH GRAPHING CALCULATORS. Assign one person in your group the role of Recorder to write your answers on a separate paper. Assign the other person the role of Grapher to use the calculator or software to do the graphs. Switch roles for each problem so that each of you gets a chance to use the calculator or software. The computer instructions are in *italic* type.

Part I

1. *Graph* the circle, $x^2 + y^2 = 16$. You will need to solve for y and enter the two equations you find. *Square* up your graph. State the center and radius of the circle.

Now we will change the equation to make the y-term negative and see what happens.

2. On the same coordinate axes as the circle above, graph the equation, $x^2 - y^2 = 16$. You will need to solve for y and graph the two equations you find. Also, remember to square up your graph. Determine at which points the graphs intersect.

The graph of the equation $x^2 - y^2 = 16$ is a **hyperbola.** The points that both the circle and hyperbola share are called the **vertices** of the hyperbola.

3. Fill in the following table using the equation of the hyperbola above (round your answers to two decimal places). Describe your observations of how the x- and y-values compare.

x	y
−10, 10	
−100, 100	
−1000, 1000	
−10,000, 10,000	

The lines that your points are approaching (getting closer to) are the **asymptotes**.

4. *Turn off* the graph of the equations for the circle and graph the hyperbola only. *Zoom* out and *square* your graph. What shape does your graph begin to take?

5. Determine the equations of the asymptotes. These should be linear equations (equations for lines).

6. Determine where the asymptotes cross. Turn the circle equation back on. How does the intersection of the asymptotes for the hyperbola relate to the circle?

The point you found in question 6 is called the **center** of the hyperbola. The standard form for each of the types of equations of a hyperbola with center (h, k) is $\frac{(x-h)^2}{a^2} - \frac{(y-k)^2}{b^2} = 1$, $a > 0$, $b > 0$ or $\frac{(y-k)^2}{b^2} - \frac{(x-h)^2}{b^2} = 1$, $a > 0$, $b > 0$. (Some books will have the a^2 always as the denominator for the positive expression and the b^2 always as the denominator for the negative expression.) To help find the asymptotes for our hyperbola algebraically, we will need the hyperbola equation from question 2 in standard form.

7. Complete the steps to put the hyperbola equation from question 2 in standard form.

The asymptotes are then: $y = \dfrac{b}{a}x$ and $y = -\dfrac{b}{a}x$.

8. What are the asymptotes for the hyperbola $\dfrac{x^2}{16} - \dfrac{y^2}{16} = 1$?

Now we will return to the issue of the vertices of a hyperbola. For this activity we will only consider hyperbolas centered at $(0,0)$.

9. Looking at your graph from question 2, notice where the circle and the hyperbola intersect. Remember that the intersection points are the vertices. How are these points related to the x and y axes?

We will explore below what would happen if the x^2 term were the negative one.

10. Return your graph to the standard axes and graph the hyperbola: $\frac{y^2}{16} - \frac{x^2}{16} = 1$ on the same coordinate axes as the hyperbola you graphed above. *Remember to square your graph.* Compare the second hyperbola and the first one. What are the asymptotes and vertices of the second hyperbola (*zoom* out if you wish to)?

11. Find a general way for how you would use the equation in standard form for a hyperbola to find the vertices without having to find a circle or ellipse equation. List the steps you would use.

Part II

12. Put the equations below in standard form and state which kind of conic each equation represents, whether a circle, ellipse, or hyperbola. Graph each pair of equations on the same coordinate axes. Remember to *square* your graphs so that the circles and ellipses will correctly appear. *Zoom* if necessary. Give the asymptotes and vertices for each hyperbola. Switch roles in your group for each pair of equations.

  a. $x^2 + y^2 = \dfrac{1}{4}$ $4y^2 - 4x^2 = 1$

  b. $4x^2 + \dfrac{y^2}{9} = 1$ $4x^2 - \dfrac{y^2}{9} = 1$

  c. $\dfrac{x^2}{16} + \dfrac{y^2}{9} = 1$ $\dfrac{y^2}{9} - \dfrac{x^2}{16} = 1$

NOTES TO THE INSTRUCTOR

Summary: Mathematical functions are related to spreadsheet functions with which students may be more familiar.

Skills Required: Some Knowledge of Spreadsheets
Skills Used: Concept of Function, Properties of Functions, Functional Notation

Grouping: Any grouping technique, make sure at least one person in each group has familiarity with spreadsheets, if possible
Materials Needed and Preparation: Copies of Activity 3-17, one for each group
Student Time: [In Class] 20 to 30 minutes; [Out of Class] none

Teaching Tips: You may want to use a computer with overhead projector to show students a sample of a spreadsheet. If the students using the activity are unfamiliar with spreadsheets, this activity could be confusing, so showing them an example and having it displayed during the activity could alleviate some of the confusion. Explain the directions in the activity to the class by going through the examples in Part I. Have students form groups or plan groups ahead of time so that each group has a member who is familiar with spreadsheets. Have students complete Parts I and II. OPTIONAL: Point out to your students that in Part II the compostion of the absolute value function and the square root function has all real numbers as its domain whereas the square root function alone only has positive real values as its domain. This can lead into a discussion of the domain and range of a function.

Grading Tips: Group Grade; 7 points for parts (a) and (b) for each "function" in Part I; 30 points for Part II.

Comments: This activity gives the students practice with the new notation for functions. Also, connecting the idea of functions with spreadsheets gives a more concrete basis for the abstraction. Many students will encounter spreadsheets in the future, either on the job or at home, if they have not already.

Spin-offs: Write and talk about some computer spreadsheet functions with which students are already familiar.

Many people in their everyday lives and in the workplace use computer spreadsheets. One of the powerful parts of a spreadsheet is the use of functions to perform tasks for you. In this activity we will relate these spreadsheet functions to mathematical functions and look at some of their properties.

A spreadsheet is made up of **cells**, which are usually labeled with a letter and number. For example, in the portion of a spreadsheet below, the cell labeled $A3$ contains the value 10.

	A	B	C	D
1				
2				
3	10			
4				

In many spreadsheets, **functions** begin with a special symbol. We will use the "@" symbol to precede the spreadsheet functions here. Below are two examples of possible spreadsheet functions. One is a true function and one is not. There is also a comparison of the spreadsheet notation to the functional notation for the real function.

EXAMPLE A: **@Sum**(*range of values*) — Adds all of the values listed, either by position in the spreadsheet or actual value, and returns the sum.

@Sum($I4 : I10$) (Suppose $I4$ to $I10$ had the values 1, 2, 3, 4, 5, 6, 7)

$I4 + I5 + I6 + I7 + I8 + I9 + I10 = 1 + 2 + 3 + 4 + 5 + 6 + 7$

Returns the value 28.

This is a function, since only one value is returned even though there are multiple values going in. In mathematics we would use the notation: $f(I4, I5, I6, I7, I8, I9, I10) = I4 + I5 + I6 + I7 + I8 + I9 + I10$.

EXAMPLE B: **@Circlepts**(r, x) — Given the radius of a circle centered at the origin, r, and an x-value, this function returns the associated y-value(s).

@Circlepts(4, 0): $x^2 + y^2 = r^2$ (equation for a circle centered at the origin)

$r = 4, x = 0$; $(0)^2 + y^2 = (4)^2$

$y^2 = 16 - 0$ (solving for y); $y^2 = 16$

$y = 4, -4$; Returns the y-values 4 and -4.

This is *not* a function, since it would try to return *two* y-values for the x-value given.

Part I
Below are listed some spreadsheet "functions." Some are true functions and some are not.

1. **@Abs**(x) — Returns the absolute value of x.

2. **@Root**(x) — Returns the value(s) that would need to be squared to get x, where x is a positive number.

3. **@Sqrt**(x) — Returns the positive square root of x.

4. **@Parab**(y, a, b, c) — Returns the x-value(s) associated with the given y-value in an equation of the form $y = ax^2 + bx + c$.

5. **@Average**(*range of values*) — Averages the values entered either by position in the spreadsheet or actual values.

For each one:

a. Give the steps the "function" will perform to return the correct value(s). Use an example to show the steps (as in Examples A and B).

b. Determine whether each one is truly a function and decide how the true ones would be written in functional notation.

Part II
These functions may also be nested inside the spreadsheet cells. For example, suppose you wished to find the square root of the absolute value of a number. In a spreadsheet you would write:

$$\textbf{@Sqrt(@Abs}(A1))$$

where $A1$ represents the position in the spreadsheet of the number with which you wish to work. In mathematics, this is called **composition of functions.** The above example would be written as:

$$f \circ g(x) = f(g(x)) = \sqrt{|x|}$$

where $f(x) = \sqrt{x}$, $x \geq 0$; $g(x) = |x|$; and $A1 = x$.

6. Using the true functions you found in Part I, write two compositions using two functions in each. Write what each would look like in a spreadsheet and what it would look like in functional notation.

NOTES TO THE INSTRUCTOR

Summary: Students use natural logarithms and exponentials to find out how long it will take to fill the Earth with people.

Skills Required: Solving Exponential Equations
Skills Used: Applications of Exponential Equations

Grouping: Proximity Pairing
Materials Needed and Preparation: Copies of Activity 3-18, one for each group
Student Time: [**In Class**] 30 minutes; [**Out of Class**] none

Teaching Tips: After covering the material on exponential growth, have the students pair up with partners to work the activity. Determine whether you want individual or group papers. You may want to allow some class time after the activity to discuss the results. Some of the questions in the activity already have some thought questions included.

Grading Tips: Group Grade; 15 points each for questions 1 through 3 and 5 through 8; 10 points for discussion participation

Comments: This activity could elicit lively discussions. Be prepared to remind students to keep the volume down and respect the opinions of others.

The formula for exponential growth is:

$$P(t) = P_0 e^{kt}$$

where $P(t)$ is the population after t years, P_0 is the population initially, k is the growth rate per year, and t is the number of years elapsed.

Part I

For all of your calculations below, find out from your instructor how he or she wants you to round your answers and carry as many decimal places as possible until your final answer.

1. If the population of the world doubles in 30 years at the current rate, what *is* the current growth rate (express as a percent)?

We are going to find out how long it will take to completely cover the Earth. The Earth has a total surface area of approximately 5.1×10^{14} square meters. Seventy percent of this surface area is rock, ice and sand, open ocean, and other locations we will consider unavailable for growing food for humans. Another 8% of the total surface area is tundra, lakes and streams, continental shelf, algal bed and reef, and estuaries, which we will consider unfit for living space.

2. a. Determine the surface area that *is* available for growing food.

    b. Determine the surface area available for living space. Notice that the surface area available for living space is also considered available for growing food.

Suppose that each person needs one square meter of the Earth's surface for living space. There are presently approximately 5.5×10^9 people on the Earth today.

3. If none of the surface area available for living space were used for food, how long would it take for the livable surface of the Earth to be covered with people (use the rate you found in question 1 and the surface area you found in question 2(b))? How much surface area would be left to grow food?

4. Measure out a space that is one square meter (a meter is slightly longer than a yard) in area and discuss with your group if you would want to be packed that closely together on the Earth. Take into account that many people live in tall apartment buildings and how that translates into surface area used per person.

Now, suppose that for each person, 100 square meters of the Earth's surface is needed for living space and growing food.

5. Using the same present population and growth rate as you did in question 3, how long will it take to cover the Earth with people? Use the surface area you found in question 2(a). Does 100 square meters per person for living space and growing food seem reasonable? How much room do you think it takes to grow animals for food (cows, chickens, pigs, etc.)? What about grains, vegetables, nuts, and fruit? Would food grow as well in desert areas, mountainous areas, or jungle areas? Would there be any space left for wild animals or natural plant life? Would there be any space left for shopping malls, movie theaters, concert halls, factories, or office buildings?

Part II

To go one more step toward reality, it is known that the growth rate has been changing (increasing) over time due to changes in technology, medicine, better food, and number of babies conceived. Population figures show that it took from 1650 to 1850 for the world population to double (200 years). But it took only from 1850 to 1950 for the population to double again (100 years), cutting the doubling time in half. This means that population is not growing at a true exponential rate.

6. Suppose the growth rate *is* exponential. Find what the growth rate would be if you halve the doubling time from question 1. Notice that decreasing the doubling time increases the growth rate. Is the growth rate doubled? Explain why or why not.

7. Redo the calculation you did in question 5 using the new growth rate you found in question 6. How much difference does this make in the time to cover the Earth (compare to your answer in question 5)?

8. Now suppose a miracle occurs which causes the present population (5.5×10^9) to double instantaneously. Using the growth rate you found in question 1, how long will it take to cover the Earth? How does this compare to halving the doubling time (compare to your answers to question 7 and question 5)? Which has the greater effect, changing the present population or changing the growth rate?

Figures show that the Earth's population has already reached 5.5×10^9 and the doubling time is decreasing (growth rate increasing) at a faster rate than the trend discussed above. Also, because of various considerations, it is not clear that the growth rate will be decreasing in the near future. Forecasters use a more accurate model of population growth which requires calculus to compute the changes in population. This model predicts the population reaching capacity within the next 100 years.

ANSWER KEYS

INTRODUCTION

Activity A: Think Teams, p. 19

Different employer surveys rate qualities in slightly different orders (refer to Activity B: *What Do Employers Want?*), but the following qualities are usually in the top three:

1. Reliable and dependable
2. Works well with others
3. Self-directed to solve problems and set goals

How this math class will help (various answers such as the following):

1. Class attendance and preparation for homework and group activities
2. Think teams and other group activities
3. Students setting goals and taking responsibility for their own learning

Activity B: What Do Employers Want? p. 21

In a survey of 186 companies, employers ranked the qualities in the order in which they are given. The survey results were published in the Lindquist-Endicott Report by Victor R. Lindquist, Northwestern University Placement Center, Evanston, Illinois.

Activity C: To the Student, p. 23

Various, names of group members

Activity D: Roles for Groups, p. 25

Various, names of group members, role assignments

Activity E: How Are Things Going? p. 28

Various

Activity F: Using College to Reach My Goals, p. 30

Various

Activity G: Group Review, p. 33

Various

Activity H: Learning from Your Mistakes, p. 36

Various

CHAPTER 1

Activity 1-1: Recycling, p. 41

1. four hundred forty-one thousand, four hundred seventy; one million, five hundred fifty-six thousand, one hundred eighty; twenty-one million   **2**. tons of paper: 441,500; 440,000; 400,000   gallons of oil: 1,556,200; 1,560,000; 1,600,000   **3**. $21 \cdot 10^6$; $5 \cdot 10^5$; $26 \cdot 10^6$ **4**. 889 lb   **5–9**. various

Activity 1-2: It's Party Time!, p. 44

1.

$1\frac{1}{2}$	small (6-oz) cans shrimp	1	dashes hot sauce
$\frac{1}{2}$	egg, boiled and mashed	$1\frac{3}{4}$	tablespoons chopped fresh chives
$\frac{1}{4}$	cup mayonnaise	$\frac{1}{8}$	teaspoon salt
$1\frac{1}{4}$	tablespoons chopped onion	$\frac{1}{4}$	teaspoon chili powder
$1\frac{5}{6}$	tablespoons plain yogurt	$\frac{1}{2}$	tablespoon lemon juice

2.

9	small (6-oz) cans shrimp	6	dashes hot sauce
3	eggs, boiled and mashed	$10\frac{1}{2}$	tablespoons chopped fresh chives
$1\frac{1}{2}$	cups mayonnaise	$\frac{3}{4}$	teaspoon salt
$7\frac{1}{2}$	tablespoons chopped onion	$1\frac{1}{2}$	teaspoons chili powder
11	tablespoons plain yogurt	3	tablespoons lemon juice

3. 48 cups of punch

4.

18	cups apples, pared and sliced	$3\frac{3}{8}$	teaspoons cinnamon
$3\frac{3}{8}$	cups brown sugar	$\frac{9}{16}$	teaspoon allspice
$2\frac{1}{4}$	cups dry oats	$1\frac{1}{8}$	teaspoons nutmeg
3	cups flour	$1\frac{1}{2}$	cups melted butter

5. 6 times   **6**. 8 batches   **7**. various   **8**. various

Activity 1-3: Stock Market, p. 47

	Stock Alpha	Stock Beta	Stock Gamma
1.			
Ending price:	$10\frac{7}{8}$	$25\frac{1}{4}$	$19\frac{9}{16}$

2–5. various

Activity 1-4: Fractions Step-by-Step, p. 50

1. $\frac{7}{8}$ **2.** $61\frac{1}{3}$ **3.** $6\frac{2}{3}$ **4.** $7\frac{5}{9}$ **5.** $3\frac{17}{21}$ **6.** $6\frac{3}{4}$ **7.** $7\frac{1}{24}$ **8.** $2\frac{11}{12}$

Activity 1-5: Do You Have Enough Money?, p. 53

9:15 A.M. **1.** $78.98 needed **11:00** A.M. **1.** $7.83 cash **2.** $47.33 bank account **noon** $14.64 cash **1:30** P.M. $6.33 bank account **2:30** P.M. $226.09 bank account **3:00** P.M. $34.64 cash, $164.52 bank account **4:00** P.M. **1.** yes **2.** yes **5:00** P.M. various

Activity 1-6: Sales Pitch, p. 56

1. Plan A: $180; Plan B: $165; Plan B is less expensive. **2.** $24.10 **3.** $41.85 **4.** $35.60 **5.** yes; various, such as, buying more CDs than usual in order to meet membership requirements, choice between buying by mail (waiting for catalogs and products) or in person (greater flexibility) **6.** $970.74 **7.** $478.35 **8.** $159.45 **9. a.** $19.99 **b.** $19.23 **10.** various, such as, Are there handling charges? Charges for shipping? Tax? Insurance? **11.** various, such as, rent to own is much more expensive but it avoids a down payment and a credit check

Activity 1-7: Making Comparisons, p. 61

Parts I and III: Answers will vary—compare answers within teams and classes for consistency. **Part II:** Presentations should cover the main points and steps needed to solve the problem(s) assigned to the team. **Page 2: 1. a.** $\frac{1}{12}$ **b.** $\frac{1}{13}$ **2. a.** yes **b.** yes **c.** no **d.** yes **3. a.** $x = 20$ **b.** $n \approx 2.3$ **4. a.** $n = 49$ **b.** $y = 10$ **5.** 0.15 miles/min **6.** the 15-oz can **7.** $\frac{3}{8}$ or 0.375 cup of sugar **8.** $7\frac{1}{2}$ or 7.5 gallons

Activity 1-8: Thinking Along These Lines, p. 64

1. D and E **2.** between B and C; 71% **3.** T and U **4.** letter V **5.** between I and J **6.** between G and H (close to G) **7.** between J and K **8.** past L (all of the line plus half again as much) **9.** 5% **10.** $2\frac{1}{2}$% **11.** $1\frac{1}{4}$% **12. a.** mark in the middle of the bottom of the page **b.** $4\frac{1}{4}$ **c.** various, such as, two $8\frac{1}{2}''$ pages or two $8\frac{1}{2}''$ lines **d.** 17 **13, 14, 15.** various diagrams **16.** 400% **17.** various

Activity 1-9: News Report, p. 67

1. 98,500 people **2.** 25% **3.** $175,000 **4.** $12,000 interest **5.** 40% decrease **6.** $6,400 commission **7.** $103.55 total cost **8.** 4,301 people **9.** $18 each **10.** $360,000 more money **11.** various

Activity 1-10: Midterm Madness, p. 70

Review Set 1: **1.** 70 **2.** 21 **3.** $\frac{6}{203}$ **4.** $x \approx 169.2$
Review Set 2: **1.** 10 **2.** 7 **3.** $n = 4.375$ **4.** 84.227
Review Set 3: **1.** $x = 88$ **2.** 20 **3.** 50 **4.** 150
Review Set 4: **1.** 72 **2.** 36% **3.** $\frac{9}{25}$ **4.** $\frac{5}{81}$
Final Challenge: **1.** 50% of the last answer from Set 1 is larger because 84.6 > 84.227 **2.** $9\frac{7}{27}$

Activity 1-11: The Survey, p. 73

1–9. various

Activity 1-12: Using Measurement, p. 76

Note: The measurements given here may be used as approximations. The actual measurements should be determined by the students and agreed upon by the class.
1. various, such as **a.** length, $10\frac{3}{4}$ in.; width, $8\frac{3}{8}$ in. **b.** length, 27.2 cm; width, 21.2 cm
2. a. length, $10\frac{1}{2}$ in.; width, 8 in. **b.** length, 26.6 cm; width, 20.3 cm **3. a.** $\frac{1}{4}$ in. **b.** 7 mm
4. a. $1\frac{5}{8}$ in. **b.** 42 mm **5. a.** 25 mm **b.** 2.5 cm **6. & 7.** various **8.** various, such as $38\frac{1}{4}$ in. **9.** various, such as 576.64 sq cm **10.** 93.8 cm **11.** 84 sq in. **12.** 13 sq in. **13.** 78.5 mm **14.** 4.90625 sq cm **15. & 16.** various

Activity 1-13: A Moving Experience, p. 79

1. 6 Small Boxes, 7 Medium Boxes, 3 Large Boxes, 2 Extra-Large Boxes, 1 Dish Pak Box, and 1 Dish Saver **2.** $47.70 **3.** approx. 61.75 cu ft, with some variations depending upon the approach used **4.** 10′ Mini Mover or 6′ × 12′ Trailer & Car Top Carrier **5.** 2 round pizzas = 226.08 sq in; rectangular pizza = 288 sq in **6.** round, approx. 5 cents per sq in; rectangular, approx. 4 cents per sq in, so rectangular is the better buy **7.** 123.75 cu ft **8.** 6′ × 12′ Trailer **9.** $84.38 **10.** various

Activity 1-14: Geometry Park, p. 83

1. a. 264 yd **b.** various **2.** 3,968 sq yd **3.** 96 sq yd **4. a.** 120 cu ft **b.** various **5.** 113.04 sq ft **6. a.** 37.68 ft **b.** various

Activity 1-15: Follow the Signs, p. 86

Part I: various **Part II: 1.** $612 **2.** 118 lb **3.** −2.8 lb **4.** 35 yard line **5.** 28 degrees **6.** 5,933 ft **7.** −2 degrees **8.** $41\frac{5}{8}$ **9.** $21 - 3 - 2(3) = 12$ ft **10.** −14 degrees

Activity 1-16: Thinking Aloud, p. 89

1. c **2.** b **3.** $3a$ **4.** $-6a^2$ **5.** $7y$ **6.** 3 **7.** a **8.** z **9.** $3x$ **10.** $2y$ **11.** $3x$
12. 2 **13.** $a + 4b = \boxed{1 + 4(6)} = 1 + 24 = 25$; correct: $a + 4b = 6 + 4(1) = 6 + 4 = 10$
14. $3(x + y) = 3(2 + 4) = 6 + \boxed{4} = 10$; correct: $3(x + y) = 3(2 + 4) = 6 + 12 = 18$
15. $5x - y = 5(3) - \boxed{2} = 15 - 2 = 13$; correct: $5x - y = 5(3) - (-2) = 15 + 2 = 17$ **16.** $\frac{8(0)}{4} =$
$\boxed{\frac{8}{4}} = 2$; correct: $\frac{8(0)}{4} = \frac{0}{4} = 0$ **17.** $x - 5\boxed{-}5 = 15\boxed{-}5$; correct: $x - 5 + 5 = 15 + 5$;
$x = 20$ **18.** $\boxed{-}\frac{8}{1} \cdot \frac{1}{8}t = -8 \cdot \boxed{-}\frac{8}{1}$; correct: $+\frac{8}{1} \cdot \frac{1}{8}t = -8 \cdot +\frac{8}{1}$; $t = -64$ **19.** $x = -4$
20. $x = -0.64$

Activity 1-17: "Algebragging," p. 92

Variables may vary. **1.** Let $m = $ Mia; $2m$ **2.** Let $x = $ what I make; $x + 1.00$ **3.** Let $d = $ Mia's distance; $7d$ **4.** Let $c = $ my charm; $\frac{1}{2}(c)$ **5.** Let $L = $ Leah's score; $L - 18$ **6.** Let $J = $ Jerry's jump; $J + 3$ **7.** Let $m = $ our money; $\frac{m}{5}$ **8.** Let $c = $ my car; $\frac{c}{3}$ **9.** $7d = 14$; $d = 2$ miles **10.** $L - 18 = 74$; $L = 92$ points **11.** $2x + 2,000 = 18,000$; $x = \$8,000$
12. $2J + 3 = 43$; $J = 20$ feet **13.** and **14.** various

Activity 1-18: Final Fling, p. 95

Review Set 1: **1.** 5 miles **2.** 2 **3.** -15 **4.** $x = 4$ **5.** 12.56 ft
Review Set 2: **1.** -10 **2.** 2 degrees **3.** 7 **4.** 84 cu in. **5.** $z = 4\frac{1}{4}$ or 4.25
Review Set 3: **1.** 314 sq cm **2.** 12 **3.** 40 ft **4.** 7 **5.** 21.98 cu ft
Review Set 4: **1.** $3x$ **2.** $x = 14$ **3.** 308 sq meters **4.** 45% **5.** 80

CHAPTER 2

Activity 2-1: Job Decision, p. 101

1. max. \$49,000; min. \$31,000 **2.** average sales: \$125,000; average pay: \$40,000 **3.** \$34,320 (using 40 hours per week, 52 weeks per year); answers may vary **4.** various **5.** $\approx$ \$40,320 (using \$500/month for insurance) **6.** various

Activity 2-2: Grams to Calories, p. 103

Various, see sample below.

Person #1

Name of food	Grams of fat (per serving)	Total calories (per serving)
Crack That Oat Bran	4 g per 1 oz	120
Wave Those Crackers	3 g per $\frac{1}{2}$ oz	70

Person #2

Name of food	Grams of fat (per serving)	Total calories (per serving)
Chocolate Chip G-bars	3.5 g per 28 g	120
Cream of Mushroom Soup	7 g per 4 oz	100

Person #3

Name of food	Grams of fat (per serving)	Total calories (per serving)
Honey's a Nut Cereal	1.5 g per 1 cup	120
Chunk Light Tuna	1 g per 2 oz	60

Step 2: Conversion Table for Grams to Calories

Name of food	Grams of fat = G	Calories from fat, $C = 9G$
Crack That Oat Bran	4	36
Wave Those Crackers	3	27
Chocolate Chip G-bars	3.5	31.5
Mushroom Soup	7	63
Honey's a Nut Cereal	1.5	13.5
Chunk Light Tuna	1	9

Step 3: Conversion Table for Percent of Calories from Fat

Name of food	Calories from fat $= C$	Total calories $= T$	Percent of calories from fat $= \dfrac{C}{T} \times 100\%$
Crack That Oat Bran	36	120	30%
Wave Those Crackers	27	70	39%
Chocolate Chip G-bars	31.5	120	26%
Mushroom Soup	63	100	63%
Honey's a Nut Cereal	13.5	120	11%
Chunk Light Tuna	9	60	15%

Step 4: Presenting the Results

Name of food	Percent of calories from fat $= \dfrac{C}{T} \times 100\%$
Mushroom Soup	63%
Wave Those Crackers	39%
Crack That Oat Bran	30%
Chocolate Chip G-bars	26%
Chunk Light Tuna	15%
Honey's a Nut Cereal	11%

Activity 2-3: Please Pass the Equation, p. 107

Various, see example in the Notes to the Instructor, p. 106.

Activity 2-4: Creating Applied Problems, p. 110

Various

Activity 2-5: The Store Manager's Dilemma, p. 112

1. $R = C + 0.7C$ or $R = 1.7C$; \$4.25 **2.** $S = R - 0.25R$ or $S = 0.75(1.7C) = 1.275C$; \$3.19
3. 27.5%; various; \$0.69; various **4.** $93\frac{1}{3}\%$

Activity 2-6: An Exponential Exploration, p. 114

Level 1: 1. y^{11} **2.** x^{-12} **3.** $-x^2$ **4.** x^{-1} **5.** $x^{10}y^6$ **6.** x^3y^{-2} **7.** $x^{18}y^4$ **8.** 0
Level 2: 1. $y^{-3}z^{-1}$ **2.** $2x^3 - x^4$ **3.** 1 **4.** $x^3y^4z^5$ **5.** $5x^2y - 7xy^2$ **6.** $\frac{z}{8y^3}$ **7.** a^7bc^3
8. $8x^9y^{12}z^{30}$

Level 3: 1. x^6y^3 **2.** $3x^3 - 12x$ **3.** x^2y^2 **4.** $144x^4y^6$ **5.** $\frac{y^6}{x^{12}}$ **6.** 1
Level 4: 1. $8x^8y^5$ **2.** $x^3z^9y^{-1}$ **3.** $-4x^3y^3$ **4.** $x^2y^2z^2$ **5.** $\frac{y^8}{x^2}$ **6.** $36x^6y^4$

Activity 2-7: Get That Factor Off My Back! p. 117

The polynomials on the activity sheet are paired with their equivalent factored form.

Activity 2-8: How High Is That Rocket?, p. 120

1. same as missing information **2.** $H = -16t^2 + 512t$, quadratic **3.** 0 sec & 32 sec, varies **4.** 8 sec, 24 sec, varies **5.** 16 sec **6.** 4,096 ft, check by factoring the equation: $4,096 = -16t^2 + 512t$, which has only one solution, 16, and must therefore be the maximum **7.** yes

Activity 2-9: Algebraic Fraction Puzzles, p. 122

The page of algebraic fractions shows the pieces in the correct order with the incorrect piece at the end.

Activity 2-10: Graphing Charades, p. 126

Various

Activity 2-11: The Rent-A-Car Deal, p. 128

These sample solutions are found using some of the suggested costs in the Notes to the Instructor.
Name of rent-a-car agency: Deluxe Rent-A-Car, Jason's Jalopies, Lars' Lemons
Cost per day: $40, $90, $25
Cost per mile: $0.40, $0.15, $0.50
Part I: $y = 0.40x + 40$, $y = 0.15x + 90$, $y = 0.50x + 25$

Part II:

x = Miles	y = Cost		x = Miles	y = Cost		x = Miles	y = Cost
200	$120		200	$120		200	$125
300	$160		300	$135		300	$175
600	$280		600	$180		600	$325

Part III:

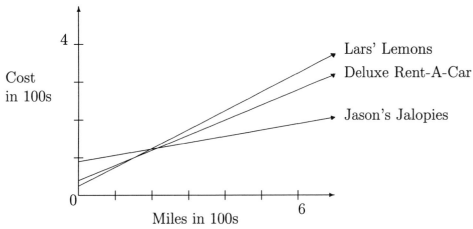

Part IV: 1. lines **2.** various, depending on equation **3.** from example: $200, $150, $225
4. See graph. **5.** various **6. a.** from example: $240, $165, $275 **b.** from example: 150, $66\frac{2}{3}$, 150
Part V: various, from example for less than 200 miles, Lars' Lemons; more than 200 miles, Jason's Jalopies

---

Activity 2-12: Celsius vs Fahrenheit, p. 131

Part I: $F = \frac{9}{5}C + 32$

Part II:

C	F
−50	**−58**
−20	−4
−17.8	0
−5	**23**
0	32

Part III:

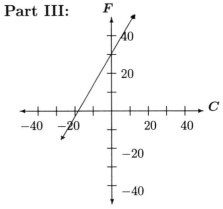

Part IV: 1. $\frac{9}{5}$ **2.** yes **3.** $(-40, -40)$ **4.** various

Activity 2-13: Interpreting Results, p. 134

Part I: **1. a.** 0 **b.** 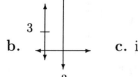 **c.** i   **2. a.** $\frac{4}{0}$ **b.** 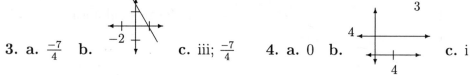 **c.** ii

3. a. $\frac{-7}{4}$ **b.** 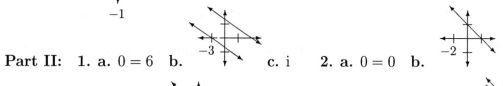 **c.** iii; $\frac{-7}{4}$ **4. a.** 0 **b.**   **c.** i

5. a. $\frac{3}{0}$ **b.** **c.** ii

Part II: **1. a.** $0 = 6$ **b.** **c.** i   **2. a.** $0 = 0$ **b.** **c.** ii

3. a. $x = 0, y = 0$ **b.** **c.** iii; $(0, 0)$ **4. a.** $0 = -1$ **b.** **c.** i

Part III:

Results	Graphical situation	Your interpretation
slope $= \frac{0}{3}$	horizontal line	**slope = 0**
slope $= \frac{5}{0}$	vertical line	**slope not defined**
$0 = 0$	equations represent the same line	**infinitely many solutions**
$0 = 3$	**parallel lines**	no solution
$x = 0$, $y = 0$	the origin	$(0, 0)$

Activity 2-14: Rent-A-Car II, p. 137

1. A: $y = 0.3x + 75$, B: $y = 0.6x + 25$
2.

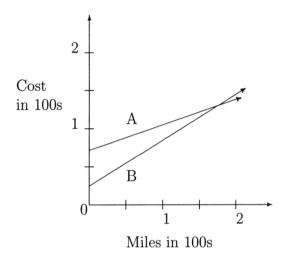

Cost in 100s vs. Miles in 100s

3. B   **4.** A   **5.** cost = \$125, $166\frac{2}{3}$ miles, 166.7, $166.\overline{6}$ **6.** A: $\{x | x > 166\frac{2}{3}\}$, B: $\{x | x < 166\frac{2}{3}\}$
7. equations; A: $y = 75(4) + 0.3(500) = 450$, B: $y = 25(4) + 0.6(500) = 400$, B is the best
8. Yes, assuming you drop the car off on Wednesday.

Activity 2-15: The Search for the Perfect Square, p. 139

Part 1: **1.** 4, 9, 16, 25, 36, 49, 64, 81, 100   **2.** $\sqrt{4} = 2$, $\sqrt{9} = 3$, $\sqrt{16} = 4$, $\sqrt{25} = 5$, $\sqrt{36} = 6$, $\sqrt{49} = 7$, $\sqrt{64} = 8$, $\sqrt{81} = 9$, $\sqrt{100} = 10$
Part II: **1.** **Student #1:** 9, 25, 49, 81   **Student #2:** 16, 36, 64, 100   **2.** various
3. various   **4.** Check work.
Part III: **1.** $4\sqrt{2}$ **2.** $10\sqrt{2}$ **3.** $7\sqrt{2}$ **4.** $5\sqrt{5}$ **5.** $9\sqrt{2}$ **6.** 8   **7.** $6\sqrt{2}$ **8.** $3\sqrt{3}$

Activity 2-16: Going a Round with Square Roots, p. 142

Round 1   **1.** 6   **2.** 3   **3.** $4\sqrt{2}$ **4.** $\frac{\sqrt{2}}{3}$
Round 2   **1.** $x^2 y^2 \sqrt{5y}$ **2.** $\frac{5}{a}$, $a \neq 0$ **3.** $9x\sqrt{x} - 2\sqrt{x}$ **4.** $\frac{3\sqrt{x}}{x}$
Round 3   **1.** $5ab\sqrt{3ab}$ **2.** $3y$ **3.** 0   **4.** $5\sqrt{5} + 10$
Round 4   **1.** $6x^5 y^2 \sqrt{2}$ **2.** $\frac{3x^3 \sqrt{6x}}{4}$ **3.** $3a^3\sqrt{3} + 3a^4\sqrt{a}$ **4.** $\frac{\sqrt{6a}}{2a}$, $a \neq 0$
Round 5   **1.** $a + \sqrt{2a} - 4$ **2.** -7 **3.** $\sqrt{x+2}$ **4.** $\frac{8 + 2\sqrt{2} - 4\sqrt{5} - \sqrt{10}}{14}$

Activity 2-17: The Maximum Playground, p. 145

1. various   **2. a.** $2W + L = 60$ **b.** $A = WL$ **c.** $A = -2W^2 + 60W$ **3.** parabola concave down, vertex at $(15, 450)$ **4. a.** the vertex $(15, 450)$ **b.** 450 sq ft   **c.** 15 ft $\times$ 30 ft

Activity 2-18: When Linear Meets Quadratic, p. 148

Answers will vary.

EXAMPLE: **Part I**   **1.** $y = 2x$; $y = \frac{1}{2}x$ **2.** $m = 2$; $m = \frac{1}{2}$ **3.** $(0,0)$; $(0,0)$ for both

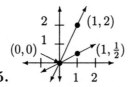

4. $(1,2)$, $(2,4)$; $(2,1)$, $(4,2)$ **5.**   **6. a.** yes   **b.** intersect at $(0,0)$

Part II   **1.** $y = 2x^2 + 2$ **2.** up, since the coefficient of x^2 is positive   **3. a.** $2x^2$, 2 **b.** 2, 2

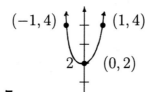

c. 2 **4.** $(0,2)$ **5.** no x-intercepts   **6.** $(0,2)$ **7.**

CHAPTER 3

Activity 3-1: A Translating Team, p. 153

Answers will vary.
Part I (examples): T: Number of people on the exploration team, $T+2$ L: Plants listed, $50\%(L)$ E: Estimated costs, $E - 30\%(E)$ S: Number of sectors, $\frac{S}{4}$ P: Number of insects predicted I: Number of insects observed, $P - I$ or $I - P$ (may be equations $= 10$) G: Supplies given, $\frac{1}{3}G$ R: Number of animals in sector three, $3R + 5$
Part II: Various, three sentences containing phrases referring to: 7 more than the number of furry creatures found   30% of supplies needed   half of what was wanted

Activity 3-2: OOOP! Order of Operations Game, p. 156

Various, see examples in Notes to the Instructor, p. 155.

Activity 3-3: Working with Sets, p. 159

Various

Activity 3-4: Building a Road, p. 161

1. 2500 ft, 0.47 mile   **2.** model of 6% grade   **3.** no, see student work   **4.** 10%   **5.** model of 10% grade   **6.** various (The Federal Highway Administration has determined the maximum safe grade for highways of this type to be not more than 8%.)   **7.** various

Activity 3-5: Parallel and Perpendicular Explorations, p. 163

Part I: **1.** See Figure 1, S: $y = \frac{-1}{3}x + \frac{11}{3}$.   **2.** See Figure 1, $(2,6)$ $(-1,7)$.   **3.** See Figure 1, P: $y = \frac{-1}{3}x + \frac{20}{3}$.   **4.** The lines have the same slope but different y-intercepts.   **5.** The lines have the same slope, different y-intercepts, and did not change the slope by moving the same number of units from both points.   **6.** $(-3,3)$ $(-6,4)$ **7.** See Figure 1, L: $y = \frac{-1}{3}x + 2$.
8. The lines have the same slope but different y-intercepts.   **9.** see Figure 1, $(1,3)$ $\left(-\frac{1}{2},4\right)$,
R: $y = \frac{-2}{3}x + \frac{11}{3}$.   **10.** R is not parallel to any of the other lines, has a different slope.
11. various

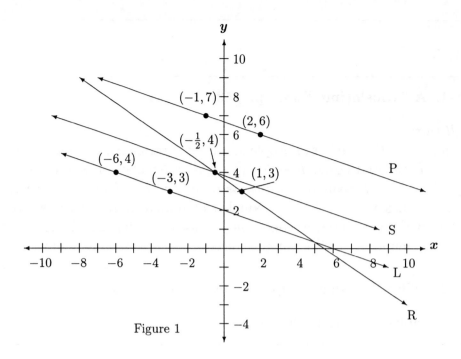

Figure 1

Part II: **12.** See Figure 2, T: $y = \frac{3}{4}x - \frac{5}{4}$. **13.** See Figure 2, $(6, -3)$. **14.** See Figure 2, Q: $y = \frac{-4}{3}x + 5$. **15.** The lines are perpendicular, the slopes are similar (product of slopes is -1 or they are negative reciprocals). **16.** The movements given in question 13 describe the slope. **17.** See Figure 2, M: $y = \frac{-4}{3}x - \frac{10}{3}$. **18.** Line M is perpendicular to line T and parallel to line Q; equations have the appropriate similarities. **19.** See Figure 2, $(2, -3)$. **20.** See Figure 2, F: $y = 4x - 11$. **21.** Line F is completely different from the others, is neither parallel nor perpendicular to any of them. **22.** Would need to move right 3 units and down 4 or left 3 and up 4 to describe the correct slope. **23.** various

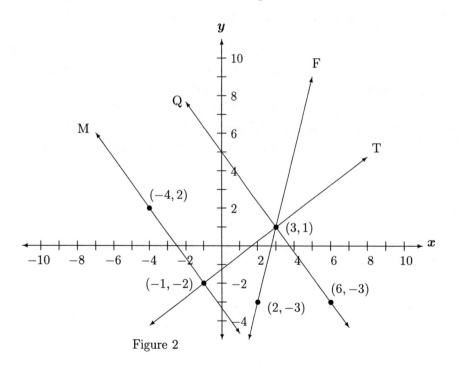

Figure 2

214

Activity 3-6: Orange Juice Demonstration, p. 166

1. 4   **2.** $\frac{1}{4} = 25\%$ **3.** water

4.

	Original juice mixture	New juice mixture
Amount of solution	4 cans	$4 + x$ cans
Amount of juice concentrate in solution	1 can	1 can

$;\quad \frac{1}{4+x} = \frac{1}{5}; \; x = 1$

5.

	Original juice mixture	20% juice mixture	New juice mixture
Amount of solution	4 cans	5 cans	9 cans
% of juice concentrate	25%	20%	x
Amount of juice concentrate in solution	1	1	2

Resulting concentration = 22.2%

Activity 3-7: Coffee on the Run!, p. 168

1. Variable names may differ: c, cups of caffeinated coffee, d, cups of decaffeinated coffee

2.

Raw material	Amt. needed for caf. per cup	Amt. needed for decaf. per cup	Max. amt. of raw material available
Bottled water (fl oz)	8	8	1280
Caf. coffee grounds (oz by weight)	0.2	0	32
Decaf. coffee grounds (oz by weight)	0	0.15	16

3. $8c + 8d \le 1280$; $0.2c \le 32$; $0.15d \le 16$ **4.** $c \ge 0$; $d \ge 0$

5.

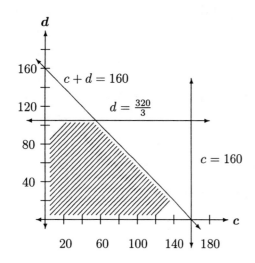

6. $8c + 8d = 1280$; $0.2c = 32$; $0.15d = 16$; $c = 0$; $d = 0$ **7.** (c, d): $(0, 0)$; $(0, \frac{320}{3})$; $(0, 160)$; $(160, 0)$; $(160, \frac{320}{3})$; $(\frac{160}{3}, \frac{320}{3})$ **8.** $(0, 160)$; $(160, \frac{320}{3})$

9.

Raw material	Cost per cup for caffeinated	Cost per cup for decaffeinated
Bottled water	$0.025	$0.025
Caf. coffee grounds	$0.028	0
Decaf. coffee grounds	0	$0.042

Cost per cup for caffeinated: $0.053; for decaffeinated: $0.067
10. Net profit per cup of coffee for caffeinated: $0.50 - 0.053 = \$0.447$; decaffeinated: $0.50 - 0.067 = \$0.433$
11. Total net profit $= 0.447c + 0.433d$
12. $(0, 0)$: Total net profit $= 0.447(0) + 0.433(0) = 0$;
$(0, \frac{320}{3})$: Total net profit $= 0.447(0) + 0.433(\frac{320}{3}) = \$46.1867 \approx \$46.19$;
$(160, 0)$: Total net profit $= 0.447(160) + 0.433(0) = \$71.52$;
$(\frac{160}{3}, \frac{320}{3})$: Total net profit $= 0.447(\frac{160}{3}) + 0.433(\frac{320}{3}) = 23.84 + 46.1867 = \$70.0267 \approx \$70.03$
The most profit comes from producing 160 cups of caffeinated and no decaffeinated. Second choice is producing about 53 cups of caffeinated and 106 cups of decaffeinated.
13. various

Activity 3-8: Building a Sunroom, p. 171

1. See figure below.   **2.** $x =$ the internal width of the sunroom   **3.** $(x + 2)(2x + 4) = 392$, $x^2 + 4x - 192 = 0$, $x = 12$ ft, $2x = 24$ ft   **4.** various   **5.** $x + 2 = 14$ ft, $2x + 4 = 28$ ft
6. various, depending on the answers to question 4. If the students use the information in the instructor notes, the answer would be yes, it will be 40 ft from the west property line and 8 ft from the south property line.

216

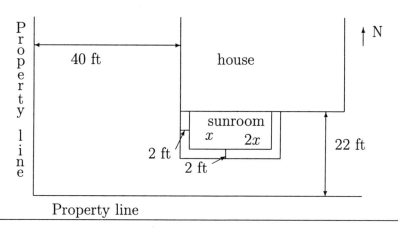

Property line

Activity 3-9: Making a Bid, p. 173

1. 12 hours **2.** 3 trees in 2 hours **3.** 5 hours **A–B,** various **4.** $(4500/A)/B = \text{rate}(r)$; where A and B are the answers to those questions asked of the landscaping company **5.** $(r + r + r + r) \cdot t_1 = 3000$ or $t_1 = \frac{3000}{4r} = \frac{750}{r}$; where r is the rate found in question 4 and t_1 is the variable **6.** $3r \cdot t_2 = 3000$ or $t_2 = \frac{3000}{3r} = \frac{1000}{r}$ **7.** All are paid the same rate, \$6.00. 5 hrs $\cdot \$6 \cdot 3 = \$90$ (three people planting the trees) $+ t_1 \cdot \$6 \cdot 4 + t_2 \cdot \$6 \cdot 3$ (putting down the sod) $=$ total cost

Activity 3-10: Water Works!, p. 175

1. 4 gal/min; 13 min **2.** 4.44 gal/min; 11.25 min **3.** 3.5 gal/min; 17.5 gal **4.** 4.5 gal/min; 22.5 gal **5.** various

Activity 3-11: Complex Numbers, p. 177

1. See Figure 1.

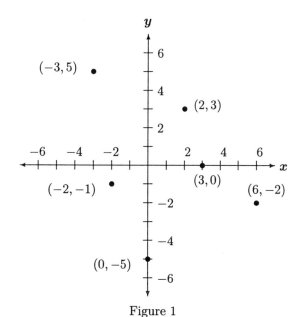

Figure 1

2. a. $2 + 3i$ **b.** $-3 + 5i$ **c.** $-2 - i$ **d.** $6 - 2i$ **e.** 3 **f.** $-5i$ **3. a.** $10 + 7i$ **b.** $-2 - 2i$ **c.** $1 - i$ **d.** 6 **e.** $-3 + 5i$ **f.** $-9i$; See Figure 2.

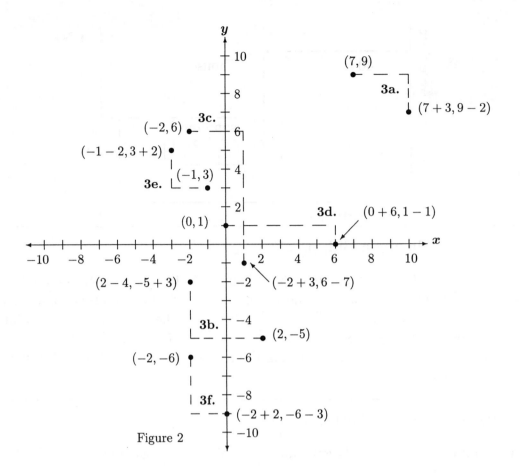

Figure 2

4.

	Complex number	Complex conjugate
a	$9i$	$-9i$
b	$3 - 4i$	$3 + 4i$
c	$-1 + 3i$	$-1 - 3i$
d	10	10

5. a. $(0,9)$; $(0,-9)$ **b.** $(3,-4)$; $(3,4)$ **c.** $(-1,3)$; $(-1,-3)$ **d.** $(10,0)$; $(10,0)$. See Figure 3. **6.** See Figure 3; description of how conjugation corresponds to reflection across the x-axis or how the lines are "mirror images."

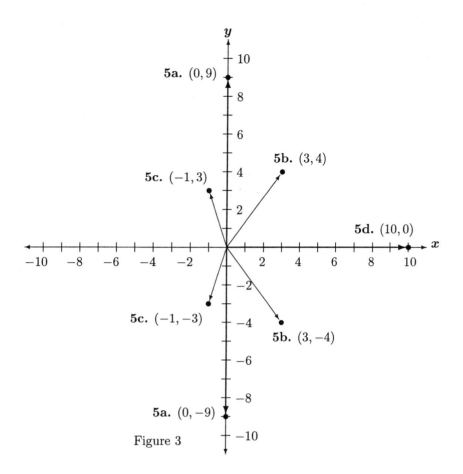

5a. $(0, 9)$

5b. $(3, 4)$

5c. $(-1, 3)$

5d. $(10, 0)$

5c. $(-1, -3)$

5b. $(3, -4)$

5a. $(0, -9)$

Figure 3

Activity 3-12: Not That Sunroom Again!, p. 180

1. assign roles **2.** r = inside radius of sunroom **3.** $\frac{1}{2}\pi\left(r^2 + 8r + 32\right) \leq 400$ sq ft **4.** $r \leq$ 11.45 ft; the south wall of the house should be greater than 34.9 ft **5.** group or individual papers

---

Activity 3-13: Getting Rational About Inequalities, p. 183

Answers will vary, see example in Notes to the Instructor, p. **182**
1. List factors from original problem. **2.** one factor per box from the original problem or question 1 with student's name next to it **3.** "break points" for each factor on the record sheet **4.** inequalities and solutions for each factor on record sheet **5.** graphs on the number line **6.** solution to original problem, pick correct interval(s) from graph.

Activity 3-14: A Carnival of Conics—Parabolas, p. 186

Part I: **1.** Graph is different from graph of lines, describe shape of parabola. **2.** $(5, 3)$
3. Graph can fold along line of symmetry, $x = 5$. **4.** vertex $(5, 3)$; x-value found in the binomial that is squared, y-value found as the constant term; line of symmetry $x = 5$; found x-value from vertex or from the binomial that is squared **5.** **a.** The new graph is wider than the original (fractional coefficient for the x-term). **b.** The new graph points down or is

flipped over (negative coefficient for the x-term).   **6.** The new graph is reflected through the $y = x$ line; vertex $(3,5)$; line of symmetry, $y = 5$; the x- and y-values have switched position.
Part II: **7. a.** $y = 3(x + 2)^2 - 4$; $(-2, -4)$; $x = -2$ **b.** $y = -\frac{1}{3}(x + 2)^2 - 4$; $(-2, -4)$; $x = -2$ **c.** $x = 3(y + 2)^2 - 4$; $(-4, -2)$; $y = -2$ **d.** $x = -3(y + 2)^2 - 4$; $(-4, -2)$; $y = -2$
8. b is flipped over the x-axis (points down) and is wider   **9.** c is reflected through the $y = x$ line, points right   **10.** d is flipped over the y-axis (points left)   **11.** c was reflected through the $y = x$ line from a; d would be reflected through the $y = -x$ line.

Activity 3-15: A Carnival of Conics—Circles and Ellipses, p. 189

Part I: **1.** various descriptions of a circle   **2.** center: $(2,5)$; horizontal and vertical lengths $= 6$ and are the same   **3.** center: $(2,5)$; found center point from the x- and y-terms in the equation in standard form; radius $= \sqrt{9} = 3$; found radius from constant term in equation   **4.** various descriptions of an ellipse   **5.** center: $(2,5)$; horizontal and vertical lengths are different: horizontal $= 4$, vertical $= 6$ **6.** center: $(2,5)$; found the center from the x- and y-terms   **7.** the numbers under the squared binomials are the squares of half of the horizontal and vertical lengths   **8.** various descriptions comparing the circle and ellipse equations; various descriptions comparing the circle and ellipse graphs   **9.** various descriptions comparing the circle and ellipse equations, such as the equation: $\frac{(x-2)^2}{9} + \frac{(y-5)^2}{9} = 1$ is the circle equation in a different form, it is in standard form for an ellipse; circles are special cases of ellipses
Part II: **10. a.** $9x^2 + y^2 = 9$; standard form: $x^2 + \frac{y^2}{9} = 1$; ellipse; center: $(0,0)$; major axis: 6; minor axis: 2; $x^2 + y^2 = 9$; in standard form; circle; center: $(0,0)$; radius: 3; diameter of circle same as major axis   **b.** $25x^2 + y^2 = 25$; standard form: $x^2 + \frac{y^2}{25} = 1$; ellipse; center: $(0,0)$; major axis: 10; minor axis: 2; $x^2 + 25y^2 = 25$; standard form: $\frac{x^2}{25} + y^2 = 1$; ellipse; center: $(0,0)$; major axis: 10; minor axis: 2; direction of major and minor axes are switched   **c.** $4x^2 + \frac{1}{9}y^2 = \frac{4}{9}$; standard form: $\frac{x^2}{\frac{1}{9}} + \frac{y^2}{4} = 1$; ellipse; center: $(0,0)$; major axis: 4; minor axis: $\frac{2}{3}$; $4x^2 + y^2 = 4$; standard form: $x^2 + \frac{y^2}{4} = 1$; ellipse; center: $(0,0)$; major axis: 4; minor axis: 2; the minor axis much smaller on the first   **d.** $9x^2 - 18x + y^2 + 4y = -4$; standard form: $(x-1)^2 + \frac{(y+2)^2}{9} = 1$; ellipse; center: $(1, -2)$; major axis: 6; minor axis: 2; $x^2 - 2x + y^2 + 4y = 4$; standard form: $(x - 1)^2 + (y + 2)^2 = 9$; circle; center: $(1, -2)$; radius: 3; diameter same as major axis

---

Activity 3-16: A Carnival of Conics—Circles, Ellipses, and Hyperbolas, p. 192

Part I: **1.** center: $(0,0)$; radius: 4   **2.** intersect at $(4, 0)$ and $(-4, 0)$
3.

x	y
$-10, 10$	± 9.16
$-100, 100$	± 99.92
$-1000, 1000$	± 999.99
$-10,000, 10,000$	$\pm 10,000$

The x- and y-values are getting to be the same or negatives of each other.   **4.** The graph begins to look like crossed lines.   **5.** $y = x$ and $y = -x$ **6.** The asymptotes cross at $(0,0)$; the same point as the center of the circle.   **7.** $\frac{x^2}{16} - \frac{y^2}{16} = \frac{16}{16}$; $\frac{x^2}{16} - \frac{y^2}{16} = 1$ **8.** $y = \pm\frac{4}{4}x$; $y = \pm x$ **9.** The points are x-intercepts.   **10.** The new hyperbola is pointing up and down rather than left and right ($90°$ rotation) but the shape is the same; the asymptotes would
220

be the same but the vertices for the new one would be $(0, 4)$ and $(0, -4)$. **11.** find the intercepts of the hyperbola equation using only the positive term (if x-term is positive then will have x-intercepts, if y-term is positive then will have y-intercepts); a list of steps to find the intercepts

Part II: **12. a.** $x^2 + y^2 = \frac{1}{4}$; already in standard form; circle; $4y^2 - 4x^2 = 1$; standard form: $\frac{y^2}{\frac{1}{4}} - \frac{x^2}{\frac{1}{4}} = 1$; hyperbola; asymptotes: $y = \pm x$; vertices: $(0, \frac{1}{2})$ and $(0, -\frac{1}{2})$ **b.** $4x^2 + \frac{y^2}{9} = 1$; standard form: $\frac{x^2}{\frac{1}{4}} + \frac{y^2}{9} = 1$; ellipse; $4x^2 - \frac{y^2}{9} = 1$; standard form: $\frac{x^2}{\frac{1}{4}} - \frac{y^2}{9} = 1$; hyperbola; asymptotes: $y = \pm 6x$; vertices: $(\frac{1}{2}, 0)$ and $(-\frac{1}{2}, 0)$ **c.** $\frac{x^2}{16} + \frac{y^2}{9} = 1$; already in standard form; ellipse; $\frac{y^2}{9} - \frac{x^2}{16} = 1$; already in standard form; hyperbola; asymptotes: $y = \pm \frac{3}{4}x$; vertices: $(0, 3)$ and $(0, -3)$

Activity 3-17: Functioning with Spreadsheets, p. 195

1. a. various, *Something Similar to:* **@Abs**(x): $|x|$, returns positive value, EXAMPLE: **@Abs**(-3): $|-3| = 3$ **b.** This one is a function. **2. a.** various, *Something Similar to:* **@Root**(x): $y^2 = x$; $y = \pm\sqrt{x}$; return values for y, EXAMPLE: **@Root**(16): $y^2 = 16$; $y = \pm\sqrt{16}$; $y = 4, -4$; **b.** This one is not a function. **3. a.** various, *Something Similar to:* **@Sqrt**(x): $\sqrt{x}$, EXAMPLE: **@Sqrt**(16): $\sqrt{16} = 4$; **b.** This one is a function. **4. a.** various, *Something Similar to:* **@Parab**(y, a, b, c): $y = ax^2 + bx + c$; $y - c = ax^2 + bx$; $\frac{y-c}{a} = x^2 + \frac{b}{a}x$; $\frac{y-c}{a} = x^2 + \frac{b}{a}x + \left(\frac{b}{2a}\right)^2 - \left(\frac{b}{2a}\right)^2$; $\frac{4a}{4a} \times \frac{y-c}{a} + \frac{b^2}{4a^2} = x^2 + \frac{b}{a}x + \left(\frac{b}{2a}\right)^2$; $\frac{4a(y-c)+b^2}{4a^2} = \left(x + \frac{b}{2a}\right)^2$; $\pm\sqrt{\frac{4a(y-c)+b^2}{4a^2}} = x + \frac{b}{2a}$; $\frac{-b \pm \sqrt{4a(y-c)+b^2}}{2a} = x$ (a more general form of the quadratic formula, notice that y is not necessarily 0). EXAMPLE: **@Parab**$(4, 1, 2, 1)$: $4 = x^2 + 2x + 1$; $3 = x^2 + 2x$; $3 = x^2 + 2x + (1)^2 - (1)^2$; $3 + 1 = x^2 + 2x + 1$; $4 = (x+1)^2$; $\pm\sqrt{4} = x + 1$; $-1 \pm \sqrt{4} = x$; $x = -3, 1$; **b.** This one is not a function. **5. a.** various, *Something Similar to:* **@Average**(*range of values*): $\dfrac{A1 + A2 + \cdots + An}{n}$ EXAMPLE: **@Average**$(1,2,3)$: $\frac{1+2+3}{3} = \frac{6}{3} = 2$; **b.** This one is a function. **6.** various, *Two of Something Similar to:* **@Sqrt(@Average**$(1,2,3)$): $\sqrt{\frac{1+2+3}{3}} = \sqrt{2}$; $f(g(x,y,z)) = \sqrt{\frac{x+y+z}{3}} = \sqrt{\frac{1+2+3}{3}} = \sqrt{2}$ where $f(x) = \sqrt{x}$, $g(x,y,z) = \frac{x+y+z}{3}$, and $x = 1, y = 2, z = 3$

Activity 3-18: How Much Space Do We Need? p. 198

Answers may vary depending upon the number of decimals retained in intermediate calculations. The answers below assume that succeeding calculations use rounded answers from previous questions and are rounded to two significant digits.

1. 2.3% **2. a.** food: $\approx 1.5 \times 10^{14}$ m^2 **b.** living space: $\approx 1.1 \times 10^{14}$ m^2 **3.** ≈ 430 years, 4.0×10^{13} m^2, discussion **4.** discussion **5.** ≈ 240 years, discussion **6.** 4.6% **7.** ≈ 120 years; half the time; yes, since growth rate was doubled time would be halved **8.** ≈ 210 years, only 30 years difference from answer to question 5 but dramatically different from answer to question 7, changing the growth rate has the more dramatic effect